Analysis & Simulation of the Deep Sea Acoustic Channel for Sensor Networks

by

Anuj Sehgal

A thesis for the conferral of a Master of Science in Smart Systems

School of Engineering and Science
Jacobs University Bremen gGmbH
Campus Ring 1
28759 Bremen
Germany

E-Mail: s.anuj@jacobs-university.de
http://www.jacobs-university.de/

Copyright

All rights reserved. This book or any portion thereof may not be reproduced or used in any manner whatsoever without the express written permission of the publisher except for the use of brief quotations in a book review or scholarly journal.
Copyright © 2013 by Anuj Sehgal.

First Printing: 2013.

ISBN 978-1-304-64636-1

www.anujsehgal.com

Abstract

Nearly 70% of our planet is composed of an aquatic environment, however, due to the lack of appropriate scientific tools and also the relative hostility of the acquatic environment, much of it still remains unexplored. With the advent of global climatic changes, a pronounced energy crisis and changing ecological habitats understanding the oceans of our planet is of vital importance. Monitoring the aquatic environment continually and effectively for oceanographic data collection, offshore exploration, efficient navigation, disaster prevention and monitoring, marine bio sciences data collection, power source exploration and maintenance can now be made possible with the deployment of underwater sensor nodes (USNs).

As in terrestrial wireless sensor networks (WSNs), usage of USNs deployed across a large area of the ocean in an underwater wireless sensor networks (UWSNs) can greatly enhance the quality of data collected within the aquatic environment. Recent advancements in unmanned underwater vehicles (UUVs) greatly extends the reach and applicability of UWSNs by enabling the integration of autonomous underwater vehicles (AUVs) acting as mobile sensor nodes (MSNs) for the purposes of underwater resource exploration and also multi-vehicle & diver coordinated collaborative exploration missions for conducting complex investigations, while also enabling autonomous navigational and location determination methodologies.

However, since radio frequency (RF) transmissions do not work underwater and optical communication is only suitable for short distances, an UWSN consists of a number of mobile and static nodes that usually communicate using the acoustic channel. Using the acoustic channel for communication causes an UWSN to contend with the issues of high transmission power requirements, rapidly changing channel characteristics, multi-path echoes, possible high ambient noise and interference, high and varying propagation delays and natural ocean currents in addition to the challenges posed by simple WSNs.

As such, in order to examine the practices used by UWSNs for successful off-shore deep sea deployments this document first analyzes the underwater channel acoustic propagation model and also looks briefly at the characteristics of the underwater transducers along with the unique effect that they pose upon sonar based communication systems. The document then goes on to exploring the state of the art in UWSNs design paradigms followed by an analysis of areas that warrant research and a discussion of the work carried out during this thesis investigation along with a conclusion highlighting the contributions it makes.

Contents

1 Introduction   1

I   Basic principles of underwater acoustics   3

2 Acoustic propagation in the ocean   4
   2.1 Speed of sound 4
   2.2 Propagation Loss 7
      2.2.1 Geometrical spreading 8
         2.2.1.1 Spherical Spreading 8
         2.2.1.2 Cylindrical Spreading 9
      2.2.2 Attenuation by absorption 9
         2.2.2.1 Absorption Mechanism 10
         2.2.2.2 Thorp Equation 11
         2.2.2.3 Fisher & Simmons Equation 11
         2.2.2.4 Ainslie & McColm Equation 13
   2.3 Transmission loss 15

II   State of the art in Underwater Networks   17

3 MAC Protocols   18
   3.1 Protocol Background 18
      3.1.1 Frequency Division Multiple Access (FDMA) 18
      3.1.2 Time Division Multiple Access (TDMA) 18
      3.1.3 Carrier Sense Multiple Access (CSMA) 19
      3.1.4 Contention-based methods (RTS/CTS, MACA, IEEE 802.11) 19
      3.1.5 Code Division Multiple Access (CDMA) 19
   3.2 Recent work 20
   3.3 Future directions 21

4 Network Topologies, Mobility and Sparsity   22
   4.1 Static Networks 23
      4.1.1 2-D Underwater Sensor Networks 23

	4.1.2 3-D Underwater Sensor Networks	24
4.2	Disruption-Tolerant Networks	25
4.3	Data and Localization Signals	26

5 Existing Evaluation Methodologies — 28
- 5.1 Simulation Environments — 28
  - 5.1.1 NS-2 Based Underwater Channel Simulator — 28
  - 5.1.2 OPNET Based Underwater Channel Simulator — 29
  - 5.1.3 MATLAB Based Underwater Channel Simulator — 30
  - 5.1.4 NetMarSys - Networked Marine Systems Simulator — 30
- 5.2 Laboratory Test-beds — 31
  - 5.2.1 Aqua-Lab — 31

III Underwater Acoustic Channel Model Development, Analysis and Simulation — 33

6 Model Development and Numerical Analysis — 34
- 6.1 The Underwater Acoustic Propagation Model — 34
  - 6.1.1 Propagation Delay — 35
  - 6.1.2 Propagation Loss — 35
  - 6.1.3 Absorption Coefficient — 36
    - 6.1.3.1 Thorp Model — 36
    - 6.1.3.2 Fisher & Simmons Model — 36
    - 6.1.3.3 Ainslie & McColm Model — 36
  - 6.1.4 Ambient Noise Model — 37
- 6.2 The Underwater Acoustic Channel Model — 37
  - 6.2.1 Received Signal Power — 38
  - 6.2.2 Signal-to-noise ratio — 38
  - 6.2.3 Optimal Transmission Frequencies — 38
  - 6.2.4 Bandwidth — 39
  - 6.2.5 Channel Capacity — 39
- 6.3 Numerical Evaluation — 40
  - 6.3.1 Optimal Frequencies — 40
  - 6.3.2 Bandwidth and Capacity — 43
  - 6.3.3 Discussion — 46

7 Software Implementation — 48
- 7.1 The AquaTools NS-2 Underwater Simulation Toolkit — 49
  - 7.1.1 Underwater Propagation Model — 50
  - 7.1.2 Underwater Channel Model — 52
  - 7.1.3 Underwater Physical Layer Model — 53
  - 7.1.4 Underwater Modulation Model — 54
- 7.2 The USARSim Wireless Simulation Server — 55

		7.2.1 Underwater Vehicle and Environment Model	55
		7.2.2 Wireless Simulation Server	56
	7.3	Discussion	59

8 Simulator Validation   61
   8.1 Noise 62
   8.2 Propagation Delay 64
   8.3 Signal-to-noise Ratio 65
   8.4 Signal Strength 68
   8.5 Bandwidth and Capacity 70
   8.6 Discussion 72

9 Conclusions & Future Directions   73
   9.1 Contributions 73
   9.2 Conclusions 74
   9.3 Future Directions 76

IV   Appendices   78

A Characteristics of Sound Velocity Parameters   79
   A.1 Ocean Temperature Profile 79
   A.2 Ocean Salinity Profile 80
       A.2.1 Salinity-Depth Profile 80
       A.2.2 Surface Salinity Profile 81

B Sound Energy Units   82
   B.1 Pascals 82
   B.2 Decibels 82

C NS-2 Sample Scripts   83
   C.1 Sample 1 - Static Nodes 83
   C.2 Sample 2 - Mobile Nodes 87
   C.3 Sample 3 - Energy Model 90

Acronyms   94

References   95

List of Figures

2.1 Speed of sound in ocean water relative to depth and water temperature (salinity fixed at 35 ppt) 5
2.2 Speed of sound in ocean water relative to salinity (depth 8 km and temperature 30°C) 6
2.3 Attenuation coefficient with varying depth and frequency 12
2.4 Attenuation (in dB/km) as a function of Depth, Temperature and Frequency (depth fixed in depicted data slice) 13
2.5 Attenuation coefficient values as predicted by the different models (Green - Fisher & Simmons; Red - Ainslie & McColm; Blue - Thorp) 14

4.1 A Typical 2-D Underwater Network [1] 23
4.2 A Typical 3-D Underwater Network [2] 25

5.1 Aqua-Lab Testbed Setup [48] 32

6.1 Optimal frequencies as predicted by the different channel models. . 40
6.2 Optmial frequencies with changing depth. 41
6.3 Optmial frequencies with changing ocean temperature. 42
6.4 Optmial frequencies with changing ocean salinity. 43
6.5 Effect of depth on available bandwidth. 44
6.6 Effect of changing temperature on bandwidth. 45
6.7 Effect of changing temperature on capacity. 46

7.1 The NS-2 channel and physical layer functional model 49
7.2 Implementation of the getTemperature and getSalinity functions which provide respective values as a function of depth according to the globally observed average thermocline and halocline. 52
7.3 Screenshot of the USARSim default model and submarine 56
7.4 Screenshot of the USARSim WSS capable of simulating underwater networks 57
7.5 Screenshot of the propagation model configuration window 58

8.1 The changing ambient noise as per changing distance which effects the optimal frequency used for noise calculation. 62

LIST OF FIGURES

8.2 The ambient noise as obtained by the simulative and analytical study conducted by Harris *et al.* while using the Thorp model [7]. . . 63

8.3 The change in propagation delay with depth of the two nodes. The propagation delay curve follows a shape similar to that of the sound velocity profile. 64

8.4 The Propagation Delay as obtained by the simulative and analytical study conducted by Harris *et al.* [7]. 65

8.5 The SNR as predicted during the study conducted by Caiti *et al.* while characterizing the underwater communication channel. (Solid lines - 1km, Dashed lines - 2km and Dotted lines - 5km; Three different cases are different operational cases with different transmission powers. Thorp model was used for the study) [55]. . . 66

8.6 The operational scenarios used in the investigation performed by Caiti et al. while characterizing the underwater acoustic channel in operational scenarios (the black dots are the transmitter and receiver pair, whereas the solid red line represents the thermocline) [55]. 67

8.7 The AN factor's relationship with the transmission frequency being utilized. The close relationship with SNR makes AN factor useful to judge performance. Only common operational frequencies are used here. (Dashed lines - 1km transmission distance, Dotted lines - 2km transmission distance & Solid lines - 5km transmission distance; Red - Thorp, Green - Fisher & Simmons, Blue/Gold - Ainslie & McColm). 67

8.8 The arriving signal strength as predicted by the Ainslie & McColm model while the distance between the transmitting and receiving nodes was varied between 4 to 180m and the transmit power is also changed. 68

8.9 The channel capacity as predicted by the Ainslie & McColm model while the distance between the transmitting and receiving nodes was varied between 4 to 180m and the transmit power is also changed. . 70

8.10 The bandwidth and capacity as predicted by the Thorp model while the distance between the transmitting and receiving nodes was varied during the study conducted by Stojanovic *et al.* [8] (Upper line is capacity). 71

A.1 Ocean water temperature with depth [54] 79
A.2 Salinity-depth profile for South Atlantic Ocean [56] 80
A.3 Average global ocean surface salinity [57] 81

Chapter 1

Introduction

Underwater sensor networks are of great importance and find applications ranging from oceanographic research, surveillance systems, navigation, offshore exploration to disaster prevention and environmental monitoring as well. Furthermore, with the globally changing climatic conditions the oceans are one of the most severely effected environments; this coupled with the need to explore deep sea offshore energy sources greatly highlights the increasing importance that underwater networks play in monitoring and exploring this environment.

The underwater channel is not conducive to using radio frequency (RF) for communication between sensor nodes as radio waves can only propagate through sea water at very low frequencies (30-300Khz) [1]. However, wireless connectivity between sensor devices can be achieved using underwater acoustic networking [1, 2, 3, 4]. Though these acoustic networks enable the use of wireless networks in a host of applications for the underwater environment, the acoustic channel access method also poses some very important challenges to achieve real-time communications in the form of limited bandwidth capacity, low battery power availability with none to little possibility of recharging and the high likelihood of network disruptions [5]. Furthermore, to achieve the largest possible area of coverage an underwater 3-dimensional sensor network is most likely to have a sparse topology [1], which leads to the transmission power required to be considerably high. As such, to maximize the lifetime of the network, obtain optimum performance and also ensure validity of data transmitted it is extremely essential to design networking schemes that are based upon utilizing the opportunities presented by the hostile deep-sea environment. This presents the unique challenge of being able to accurately model the underwater acoustic communication channel by taking into account issues such as long and varying propagation delays, multi-path echoes, high and varying ambient noise.

Designing, implementing, using and maintaining underwater sensor networks is a very costly affair [6] making it important to be able to quickly model and evaluate these networks and their associated protocols or methodologies without the need for physical deployment. This highlights the need for simulators and test-bed environments that are accurately able to model the underwater channel environment, thereyby providing an accurate tool to researchers to rapidly prototype, design and test their underwater networks, protocols or devices without the associated exhorbitant costs.

In short, in order to be able to design efficient underwater networks that reduce transmission power, improve network throughput and provide a long network lifetime in a rapidly changing environment, it is highly important to accurately model the channel in order to perform evaluations without the need for offshore testing. As part of this proposed thesis work, the underwater acoustic channel will be analysed, some of the existing state-of-the art techniques for applications of underwater networks discussed along with comparisons of existing evaluation methodologies and test-beds. Part 1 of this document presents the basic principles governing underwater acoustics that have a pronounced effect upon underwater networks. In part 2 the document moves on to discussing some of the state of the art in underwater networks and their evaluation techniques along with a brief discussion on open issues and in part 3 details of the main investigation of the thesis work are provided.

Part I

Basic principles of underwater acoustics

Chapter 2

Acoustic propagation in the ocean

The authors of [1, 2] show us that acoustic underwater networks have far reaching applications in UWSNs and multi-AUV cooperative missions. These applications range from simple monitoring and data gathering missions to possible exploration, deployment and rescue work as well; thereby, highlighting the importance of underwater acoustic networks. Despite this relative importance of acoustic networks, and the existing interest in ocean monitoring and exploration over the years, only recently considerable interest in developing networking technologies for the underwater acoustic medium has been expressed by researchers [7], thereby leaving the area of UWSNs open for investigation.

Even though wireless connectivity is achievable underwater when using the acoustic medium for inter-device networking, the acoustic channel is considerably different from the commonly used RF channel [8]. The ocean being a highly complex system medium for the propagation of sound, due to inhomogeneities and random fluctuations, including effects of the rough seas and ocean bottom variances, warrants the creation of a robust channel model that takes into account parameters like propagation loss, ambient noise, propagation delay and bandwidth and necessary transmission power in order to construct an accurate propagation model that can be used as a basis for any evaluation of acoustic networks. As such, in order to establish a basic evaluation model for any further work, this chapter is devoted to describing in detail the basic principles governing acoustic propagation in the ocean.

2.1 Speed of sound

The prime method of wireless data communication underwater is dependent on the acoustic medium and the most basic property effecting the data-rate achievable, quality of service, latency and other important network factors in this channel is the speed of sound. Owing to the possibly rapidly changing conditions of the ocean, in order to develop a sound velocity profile with some degree of accuracy, the ocean is considered to be a stratified and range independent medium that vaires only with depth. Though it is enough to make this assumption for many ocean regions, local parameters need to be measures especially in areas of high turbidity and those containing a variety of water types (typically the thermocline, halocline and coastal regions); the information presented in this section models the ocean based upon these assumptions.

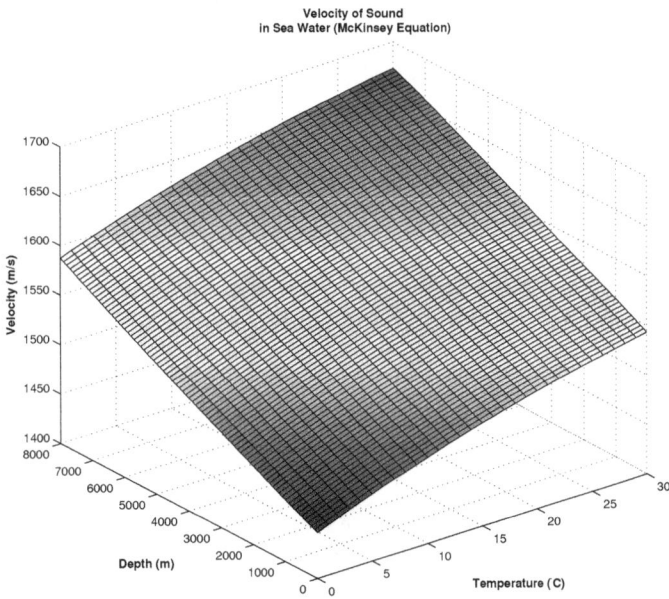

Figure 2.1: Speed of sound in ocean water relative to depth and water temperature (salinity fixed at 35 ppt)

For most purposes the speed of sound in water is taken to be approximately 1500 m/s, while this is accurate within a certain range, as it is shown in Appendix 'A', the underwater channel is an extremely complex environment that is effected by many varying factors, primarily temperature, salinity, and depth [9] and furthermore each of these factors may also be interdependent or varying across the ocean across multiple locations and depths. It is, as such, important to have an accurate model of the effects of these parameters on the speed of sound in water.

The speed of sound in water has been a focus of analysis by many mathematical models [9, 10, 11, 12, 13]. In [12] a simplified equation for the speed of sound is provided, however, after a thorough discussion of the factors effecting the speed of sound in water, the authors of [9, 11] present an expanded equation, commonly known as the MacKenzie equation (2.1), which calculates the speed of sound in water with an error in the speed estimate in the range of approximately 0.070

m/s.

$$\begin{aligned}v = &\ 1448.96 + 4.591 \cdot C - 5.304 \times 10^{-2} \cdot C^2 + 2.374 \times 10^{-4} \cdot C^3 \\ &+ 1.340 \cdot (S - 35) + 1.630 \times 10^{-2} \cdot D + 1.675 \times 10^{-7} \cdot D^2 \\ &- 1.025 \times 10^{-2} \cdot C \cdot (S - 35) - 7.139 \times 10^{-13} \cdot C \cdot D^3\end{aligned} \quad (2.1)$$

$v =$ sound speed in m/s
$C =$ temperature in degrees celsius
$S =$ salinity in parts per trillion (ppt)
$D =$ depth in meteres

Unlike the Medwin equation presented in [12] the MacKenzie equation is far more generally applicable since it does not suffer from the limitation of only being applicable up to a depth of 1 km, like its Medwin cousin. This makes the MacKenzie equation 2.1 a much better choice to be used in mathematical model developments of the speed of sound in oceans.

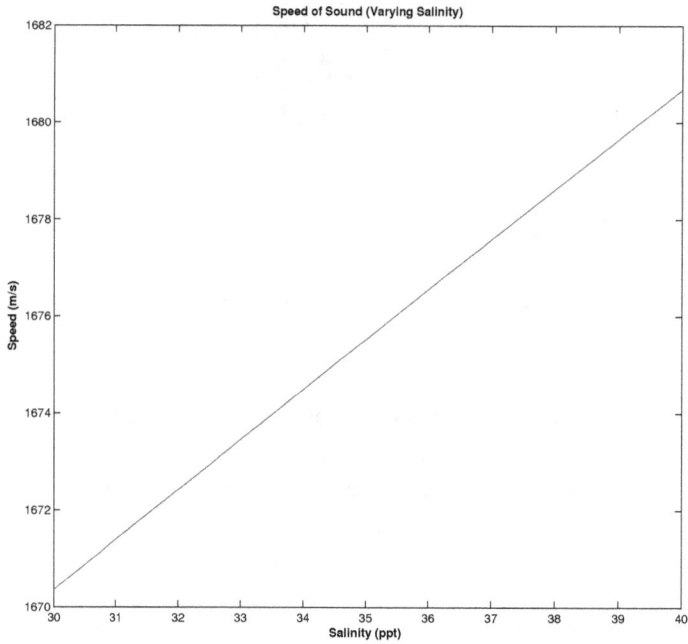

Figure 2.2: Speed of sound in ocean water relative to salinity (depth 8 km and temperature 30°C)

It is shown in Appendix 'A' that the salinty value for the ocean vaires between 30 ppt to 40 ppt, with a global depth and surface average of approximately 35 ppt. Furthermore, Figure 2.2 shows that even though the speed of sound varies with change in salinity, even at the values of temperature and depth that provide the maximum opportunity for change in speed of sound, the variance of speed over a range of 10 ppt for salinity is only about 10 m/s, thereby making the effect of changing salinity neglegible and acceptable for a constant value to be used.

Using the MacKenzie equation 2.1 a graph of the speed of sound in water, with varying depth and temperature, is plotted in Figure 2.1. The salinity in this graph is set to a value of 35 ppt in order to best display the effects of depth and temperature, the two most varying variables in a deep-sea environment. In Figure 2.1 the color of the plotted graph represents the intensity of the speed value, from blue to red represents an increase in the speed of sound.

It is clear from the graph in Figure 2.1 that the speed of sound in water is not a constant of 1500 m/s but rather varies within a range of $1400 \leq v \leq 1700$, for depths up to 8 km and temperatures up to 30°C. Furthermore, Figure 2.1 also makes it clear that the speed of sound increases with depth and also with ambient temperature; while the vertical gradient of sound velocity appears to be much larger compared to the horizontal gradient.

Sensors in an UWSN can be distributed across multiple depths, thereby encountering a range of temperatures as well. As such, both these results make it important to factor in the actual speed of sound in the environment in order to obtain an accurate result of the effects of the speed of sound on the performance of an acoustic network in deep sea environments.

2.2 Propagation Loss

The transmitted acoustic signal between sensor nodes in a network reduces in overall signal strength over distance due to a host of factors governing the sound propagation factors in ocean. This decrease of acoustic intensity between the source and receiver, termed propagation loss, is composed majorly of three aspects, namely, geometrical spreading, attenuation and the anomaly of propagation.

Geometrical spreading deals with the signal losses that occur due to focusing and defocusing effects caused by spreading of acoustic waves in the ocean water as a result of refraction, reflection and other phenomenon [14]. Attenuation is the signal loss associated with frequency dependent absorption in the underwater channel and multiple models exist to estimate the signal attenuation in ocean water. The prominent models for signal attenuation along with a discussion on the same is provided within this section.

Unlike the geometric spreading and signal attenuation, anomaly of sound propagation is extremely difficult to estimate since it encompasses all losses that might occur due to leaky communication ducts, scattering and diffraction effects that are not already attributed to geometrical spreading or attenuation. Mostly, this requires knowledge of the operation environment, however, its effects are minimalized in deep-sea areas and are mostly pronounced only in the thermocline and halocline regions [14, 15].

The overall propagation loss intensity can be calculated as a function of the acoustic intensity at the source I_s and range $r_0 \simeq 0\text{m}$ with respect to the intensity I at a range r. Authors of [16] give us a relationship for the calculation of the attenuation as a function of range and frequency, such that,

$$h(r,f) = \frac{I_s}{I} \qquad (2.2)$$

Since propagation loss consists of geometrical spreading, attenuation and the anomaly, equation 2.2 can be substituted to become:

$$h(r,f) = g(r) \cdot d(r,f) \cdot A \qquad (2.3)$$

$g(r)$, geometrical spreading of the acoustic intensity
$d(r,f)$, frequency dependent attenuation by absorption
A, anomaly of acoustic propagation

2.2.1 Geometrical spreading

Geometrical spreading of a signal comes into effect when the acoustic intensity decreases exponentially with a certain range. Spherical spreading normally occurs when the transmission distance is generally larger; on the other hand, cylindrical spreading is common in short range underwater acoustic communications. In the deep-sea sound channel a transition from the cylindrical to spherical transition also occurs [14, 15] such that if the range r is used between the sender and receiver, and r_N represents the transition range then [14],

$$\begin{aligned} r < 2\text{km} &\;:\; r_N = 1000\text{m} \\ 2\text{km} \leq r < 10\text{km} &\;:\; r_N = 1200\text{m} \\ r \geq 10\text{km} &\;:\; r_N = 5000\text{m} \end{aligned} \qquad (2.4)$$

2.2.1.1 Spherical Spreading

We know that in a homogenous and infinitely extended medium the acoustic power generated by a source gets radiated uniformly leading to a spherical spreading. The intensity at ranges r and r_0 can be represented as,

$$I_s = \frac{P_a}{4\pi r_0^2} \qquad I = \frac{P_a}{4\pi r^2}$$

r_0, reference distance ($\simeq 0\text{m}$)
P_a, acoustic power of source
I_s, acoustic intensity of source at r_0
I, acoustic intensity of source at r

As such, we get that, for spherical spreading,

$$g(r) = \left(\frac{I_s}{I}\right) = \left(\frac{r}{r_0}\right)^2 \tag{2.5}$$

2.2.1.2 Cylindrical Spreading

When the medium is confined by two reflecting panes or a shorter distance exists between the two cylindrical spreading occurs, the intensity can be represented as,

$$I_s = \frac{P_a}{2\pi h r_0} \quad I = \frac{P_a}{2\pi h r}$$

As such, we get that, for cylindrical spreading,

$$g(r) = \left(\frac{I_s}{I}\right) = \left(\frac{r}{r_0}\right) \tag{2.6}$$

2.2.2 Attenuation by absorption

Attenuation by absorption occurs due to the conversion of acoustic energy within sea-water into heat. This process of attenuation of absorption is frequency dependent since at higher frequencies more energy is absorbed. There are several equations describing the processes of acoustic absorption in seawater which have laid the foundation for current knowledge. Each of these equations has over time improved the applicability and accuracy of mathematically predicting the absorption of sound in sea water.

At low frequencies, the absorption in standard seawater is so small that immense quantities of such water are required to create measurable losses of sound energy into heat and as such the existing models may not be enough to calculate accurately the results for low frequencies.

The work of W. H. Thorp [17, 18], published in 1967, presented a simple equation to calculate the attenuation coefficient in dB/km. Through their work Fisher & Simmons [19] presented a new equation for determining attenuation coefficient by taking into account the frequency, temperature and pressure; this work was further enhanced with a new equation presented by Ainslie and McColm [20] in 1998 by also taking into account the salinity and acidity of the environment. To understand the effect of all these parameters used in these models an understanding of the absorption mechanism is required. As such, this section looks at the mechanism of absorption and then analyses the different mathematical models.

2.2.2.1 Absorption Mechanism

1. Absorption generated by particle motion

   For frequencies above 100 kHz, the particle motions generated by the sound produces heat via viscous drag. The absorption converts a proportion of the vibrational energy into heat as it travels through each successive specified distance. This proportional loss gives an exponential decay which can be specified by a ratio, or more usually by the logarithm of this ratio presented in decibels. So the results for the absorption coefficient α are usually given in dB/km for the results of measurements of attenuation at sea. An absorption of 1 dB/km means that the energy is reduced by 21 % in each successive kilometre.

   The coefficient α is found to increase with the square of the frequency f, so at frequencies greater than 1 MHz, results are usually given in dB/m, since the sound levels fall so rapidly. The value of α depends on the sea temperature T (in °C) and the pressure or depth. Whilst the conversion between pressure and depth itself depends somewhat on other parameters, these effects are small compared with the overall errors and so the use of depth D in metres is often used for convenience to calculate the hydrostatic pressure.

2. Chemical absorption

   Some molecules within sea water have more than one stable state, and changes from one to another are dependent on pressure. These changes can convert the energy associated with the fluctuating acoustic pressure into heat. Different phase changes involve different reaction times, and this lag in the response can be characterised by a relaxation time or relaxation frequency. Much faster changes have little effect as the molecular changes are too slow, so these absorption terms only affect lower frequencies [14]. Since the salinity of sea water is not the only cause for chemical absorption, the two major sources of such relaxation frequencies in the ocean are boric acid and magnesium sulphate. Please note that this document uses the nomenclature of f_1 to describe the relaxation frequency introduced by boric acid and f_2 for the relaxation frequency introduced by magnesium sulphate.

   The other parameter which has an effect on the amount of absorption in sea water is the acidity value represented by pH. Typically pH = 8 is used as the standard to represent the acidity levels of sea water. All oceans are somewhat alkaline with pH > 7, although there are concerns that this is being changed by the absorption of the excess atmospheric carbon dioxide associated with global warming [14, 15, 20].

2.2.2.2 Thorp Equation

The Thorp equation for attenuation by absorption is the simplest equation since it only takes into consideration the effect of the frequency utilized and ignores the effect of relaxation frequencies, salinity and acidity levels of the ocean.

$$\alpha = \frac{0.1f^2}{1+f^2} + \frac{40f^2}{4100+f^2} + 2.75 \times 10^{-4} \cdot f^2 + 0.003 \tag{2.7}$$

The Thorp equation shown in equation 2.7 is only applicable for a temperature of 4°C and a depth of approximately 1000m [17]. These limitations make this equation extremely difficult to be utilized in general applications of UWSNs and furthermore, by ignoring the effect of chemical absorption the equation may not necessarily produce accurate results. While this model can be used to quickly estimate the attenuation coefficient, the resulting values most likely would not be enough to produce an accurate assesment of network performance.

2.2.2.3 Fisher & Simmons Equation

The Fisher & Simmons model proposed in 1977 is one of the most commonly used and referenced models [14, 19, 15], and prior to the Ainslie & McColm equation remained the most recent one as well, thereby making it a good choice for basing most evaluations upon. Furthermore, it takes into account the effect of temperature and depth as well, while also introducing the effects of relaxation frequencies caused by boric acid and magnesium sulphate.

$$\alpha = A_1 P_1 \frac{f_1 f^2}{f_1^2 + f^2} + A_2 P_2 \frac{f_2 f^2}{f_2^2 + f^2} + A_3 P_3 f^2 \tag{2.8}$$

Equation 2.8 shows the Fisher & Simmons equation, where A_1, A_2, A_3 are functions of temperature and P_1, P_2, P_3 are functions of the constant equilibrium pressure. These are represented as:

$$
\begin{aligned}
A_1 &= 1.03 \times 10^{-8} + 2.36 \times 10^{-10} \cdot T - 5.22 \times 10^{-12} \cdot T^2 \\
A_2 &= 5.62 \times 10^{-8} + 7.52 \times 10^{-10} \cdot T \\
A_3 &= [55.9 - 2.37 \cdot T + 4.77 \times 10^{-2} \cdot T^2 - 3.48 \times 10^{-4} \cdot T^3] \cdot 10^{-15} \\
f_1 &= 1.32 \times 10^3 (T + 273.1) e^{\frac{-1700}{T+273.1}} \\
f_2 &= 1.55 \times 10^7 (T + 273.1) e^{\frac{-3052}{T+273.1}} \\
P_1 &= 1 \\
P_2 &= 1 - 10.3 \times 10^{-4} \cdot P + 3.7 \times 10^{-7} \cdot P^2 \\
P_3 &= 1 - 3.84 \times 10^{-4} \cdot P + 7.57 \times 10^{-8} \cdot P^2
\end{aligned}
$$

The values of P are represented in atm (the relationship between P and depth in meters is $P = D/10$) and f_1, f_2 are represented in Hz.

CHAPTER 2. ACOUSTIC PROPAGATION IN THE OCEAN   12

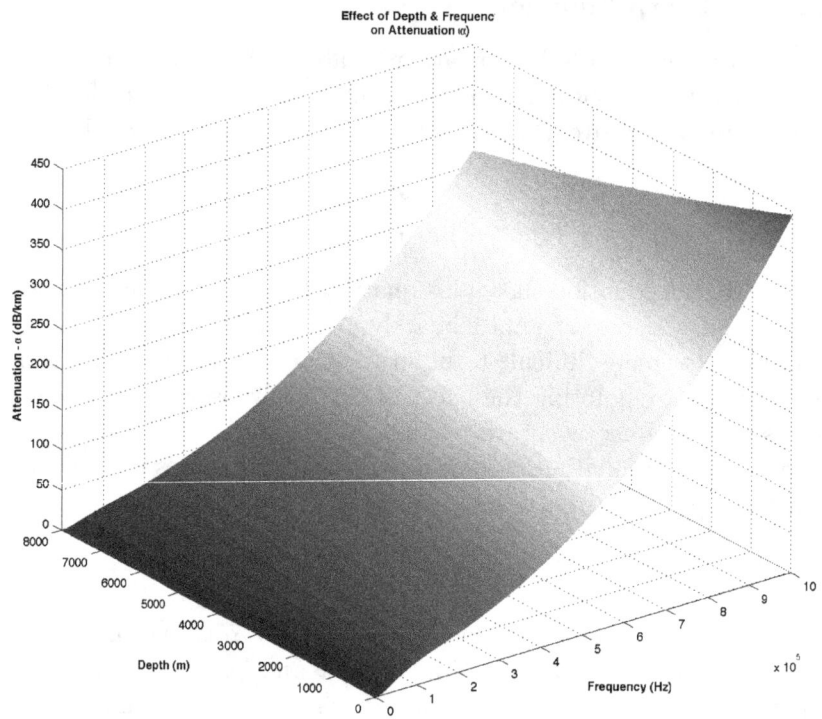

Figure 2.3: Attenuation coefficient with varying depth and frequency

The Fisher & Simmons model operates under the restriction that the depth cannot be greater than 8 km and the salinity has been restricted to a value of 35 ppt, while the pH value has been set to 8, as are the observed averages across the global ocean waters.

Using the Fisher & Simmons model, equation 2.8, we obtain the graph depicted in Figure 2.3. Though the Fisher & Simmons model is capable of calculating the coefficient of attenuation with respect to temperature as well, for the purpose of this graph the temperature was set to a value of 17°C, which has been observed to be near the global average as shown in Appendix 'A'. Figure 2.3 leads us to believe that for the attenuation constant does not increase linearly for increasing frequencies. Furthermore, the increasing depth also causes the attenuation constant to increase but with a very slight gradient.

In order to get an indication of the effect of temperature as well on the attenuation constant, Figure 2.4 presents a slice of a 4-dimensional plot of the attenuation constant with respect to the frequency, depth and temperature. Fixing the depth at 2 km in this slice shows us that with increasing temperature the value of the attenuation constant (depicted by the color) also increases.

CHAPTER 2. ACOUSTIC PROPAGATION IN THE OCEAN

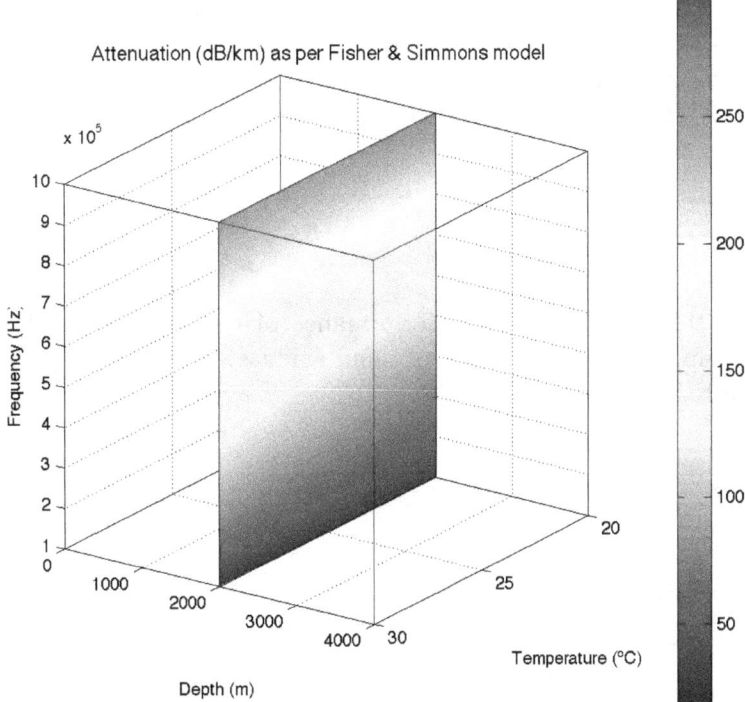

Figure 2.4: Attenuation (in dB/km) as a function of Depth, Temperature and Frequency (depth fixed in depicted data slice)

These results highlight the importance of using a model that takes into account the depth and temperature as well, when evaluating and calculating the attenuation constant that would effect the performance of an UWSN.

2.2.2.4 Ainslie & McColm Equation

The Ainslie & McColm equation proposed in 1998 is based upon the Fisher & Simmons model, however, it proposes some extra relaxations and simplifications to derive the following equation:

$$\begin{aligned}
\alpha = &\ 0.106 \frac{f_1 f^2}{f_1^2 + f^2} e^{\frac{pH-8}{0.56}} \\
&+ 0.52 \left(1 + \frac{T}{43}\right) \left(\frac{S}{35}\right) \frac{f_2 f^2}{f_2^2 + f^2} e^{\frac{-D}{6}} \\
&+ 4.9 \times 10^{-4} f^2 e^{-\left(\frac{T}{27} + \frac{D}{17}\right)}
\end{aligned} \quad (2.9)$$

Depicted in equation 2.9, the Ainslie & McColm model also takes into account the effects of the acidity of sea water and unlike the Fisher & Simmons model is based on depth (not pressure). These changes in the equation allow for a wider

CHAPTER 2. ACOUSTIC PROPAGATION IN THE OCEAN 14

range of applicability of the equation and the possibility of yielding more accurate results as well. Unlike the Fisher & Simmons model, the equations for f_1 and f_2 are also simplified and represented in kHz:

$$f_1 = 0.78\sqrt{\frac{S}{35}}e^{\frac{T}{26}}$$
$$f_2 = 42e^{\frac{T}{17}}$$

To test the comparitive performance of equations 2.7, 2.8 and 2.9 a graph with temperature, depth, salinity and acidity levels fixed to standard values[1] was generated in Figure 2.4.

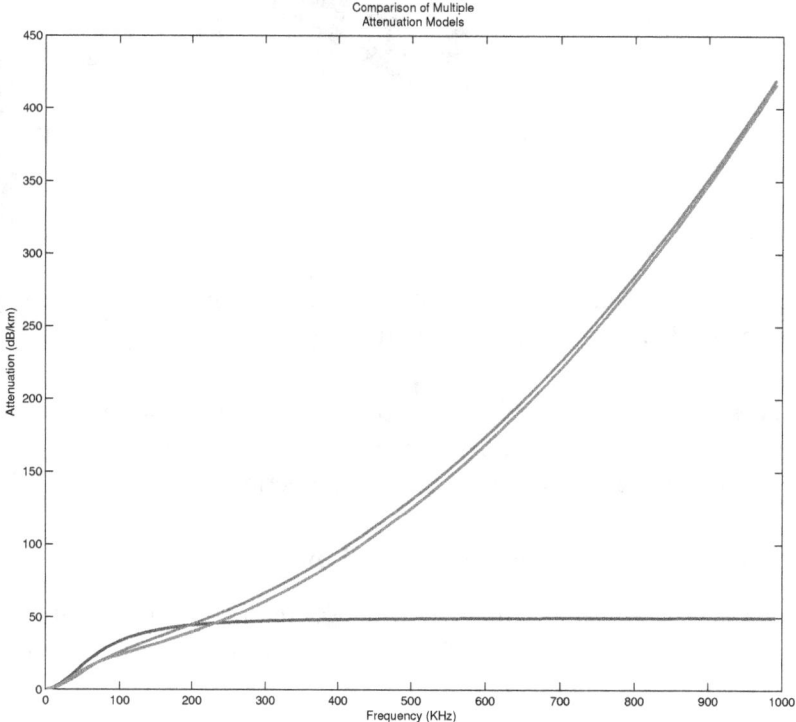

Figure 2.5: Attenuation coefficient values as predicted by the different models (Green - Fisher & Simmons; Red - Ainslie & McColm; Blue - Thorp)

It is clear from the graph that the Fisher & Simmons model and the Ainslie & McColm model have similar performance in predicting the attenuation coefficient,

---

[1] Values were picked based on the capabilities of the Thorp model and also the global observed averages, $T = 4°C$, $D = 1000m$, $S = 35$ ppt and $pH = 8$.

CHAPTER 2. ACOUSTIC PROPAGATION IN THE OCEAN   15

however, the Thorp model stops function after about a frequency of 200 kHz. This shortcoming coupled with the fact that it is restricted to a particular depth and temperature value, make the Thorp model quite unsuitable for evaluating the performance of UWSNs.

2.3 Transmission loss

Transmission loss, TL, when expressed as a single number summarizes the effect of all the aforementioned phenomenon on acoustic propagation in the sea. This TL value describes in dB the weakening of sound between two points. The TL value can be useful in determining the arriving signal strength of a data stream and even the minimum required signal strength that is necessary to successfully complete a transmission within an underwater acoustic network. TL can generally be represented by,

$$TL = 10 \log \frac{I_s}{I} \qquad (2.10)$$

In order to calculate the transmission loss that occurs due to geometrical spreading extrapolating from equations 2.5 and 2.6 into equation 2.10 we obtain the resulting transmission loss to be,

$$TL_{geometric} = 10 \log \left(\frac{r}{r_0}\right)^n = 10 \cdot n \log \left(\frac{r}{r_0}\right) \qquad (2.11)$$

where n depends upon the type of geometrical spreading that occurs. In case of cylindrical spreading, $n = 1$, whereas for spherical spreading $n = 2$.

The transmission loss that occurs due to attenuation by absorption can be calculated by the equation,

$$TL_{absorption} = \alpha \cdot \frac{r}{1000} \qquad (2.12)$$

As mentioned before, the acoustic anomaly is nearly impossible to model and as such the overall tramsmission loss occuring in ocean acoustic networks can be represented as,
$$TL = TL_{geometric} + TL_{absorption}$$
Substituting equations 2.11 and 2.12 into this relationship gives us the overall transmission loss that occurs across two nodes in a network,

$$TL = 10 \cdot n \log \left(\frac{r}{r_0}\right) + \alpha \cdot \frac{r}{1000} \qquad (2.13)$$

The transmission loss calculated by equation 2.13, though uses a value for n to take into account the effect of spherical or geometrical spreading, it does not take into account the effect of transmission loss as a result of the transition range between spherical and cylindrical spreading. This equation can be extended and simplified to the following in order to obtain the total transmission loss while also taking into account the effect of the transient range between spherical and cylindrical spreading,

$$TL = 10 \log r_N + 10 \log r + \alpha \cdot \frac{r}{1000} \qquad (2.14)$$

Equation 2.14 provides us with the total transmission loss in dB/km.

Part II

State of the art in Underwater Networks

Chapter 3

MAC Protocols

Even though media access control (MAC) has been a subject of rigorous examination for traditional radio networks and also in the case of WSNs [6, 21], it still remains an area that is largely unexplored in case of underwater acoustic networks and thereby presents a plethora of unresolved problems [1, 22, 23, 24].

Many MAC protocols have been explored for use in underwater acoustic networks, however, CDMA appears to be the most robust solution available due to its tolerance for the unique challenges presented by the underwater acoustic medium in the forms of limited bandwidth and the high and variable propagation delays. This chapter provides a little background on the advantages and shortcomings of the common MAC protocols and then looks at some of the recent work that has been carried out towards MAC protocols in the underwater acoustic channel and also outlines some of the future directions researchers are adopting.

3.1 Protocol Background

3.1.1 Frequency Division Multiple Access (FDMA)

Due to the narrow bandwidth of underwater acoustic channels and also the vulnerability of limited band systems to fading and multipath echoes FDMA is not suitable for applications in underwater acoustic networks.

3.1.2 Time Division Multiple Access (TDMA)

The long time guards required by the underwater acoustic channel lead to a limited bandwidth efficiency if TDMA is used. These long time guards are essential in the underwater acoustic medium to account for the large propagation delay and delay variance of the underwater channel and minimize packet collisions from adjacent time slots. The existence of a variable delay in the channels makes it

CHAPTER 3. MAC PROTOCOLS   19

difficult to achieve a precise synchronization with a common timing reference; this synchronization is necessary for TDMA to function.

3.1.3 Carrier Sense Multiple Access (CSMA)

Usage of the CSMA protocol prevents collisions with ongoing transmissions at the transmitter, however, to avoid collisions at the receiver, it is necessary to add a guard time between transmissions which is dimensioned proportionate to the maximum propagation delay that could exist in the underwater network. Having such a large guard time makes CSMA extremely inefficient for applications in underwater acoustic networks.

3.1.4 Contention-based methods (RTS/CTS, MACA, IEEE 802.11)

Contention-based methods relying on handshake mechanisms, such as RTS/CTS, MACA and IEEE 802.11, are not suitable for applications in the underwater acoustic channel because:

1. Large and variable propagation delays of the RTS/CTS packets can lead to a low throughput.

2. The high propagation delay characteristic of underwater acoustic channels can lead to the channel being sensed as idle, in case of carrier sense protocols like 802.11, even though a transmission might be ongoing as the signal may not have reached the receiver.

3. The high variability of delays in propagation of control packets makes it impossible to predict the start and end times of transmissions for other nodes, thereby making collisions highly likely.

3.1.5 Code Division Multiple Access (CDMA)

Since CDMA distinguishes simultaneous signals transmitted by multiple devices by using pseudo-noise codes for spreading the user signal over the entire available band, it is robust to the frequency selective fading that occurs due to multi-path propagation in underwater networks. By using Rake filters [25] designed to match the pulse spreading, shape and channel impulse response the time diversity of the underwater acoustic channel can be leveraged to correct for the effects of multi-path propagation [2].

Power efficiency is an important factor in the design of any underwater network as the available battery power to cost ratio is quite high. In this regard as well, the usage of CDMA results in decreased battery consumption and a high throughput as it allows for reducing the total number of packet transmissions. The authors

CHAPTER 3. MAC PROTOCOLS

of [26] compare two CDMA techniques, direct sequence spread spectrum (DSSS) and frequency hopping spread spectrum (FHSS) for shallow water communication. The results of this study show that FHSS is prone to Doppler shift since all transmissions occur in narrow bands, but it is more robust to multiple access interference as compared to DSSS. Their investigations also result in conclusions that even though FHSS leads to a higher bit error rate, the receivers built for it are simpler and thus simplify power efficiency control.

A new scheme presented in [27] combines multi-carrier transmission with DSSS CDMA since it offers higher spectral efficiency than the single carrier counterparts in the underwater acoustic channel. The proposed idea spreads each data symbol in the frequency domain by transmitting all the chips of a spread symbol at the same time into many narrow sub-channels in order to achieve high data rate by increasing the duration of each symbol to reduce inter-symbol interference.

One of the most attractive access techniques in the recent underwater literature combines multi carrier transmission with the DSSS CDMA [28], as it may offer higher spectral efficiency than its single carrier counterpart, and increase the flexibility to support integrated high data rate applications with different quality of service requirements. The main idea is to spread each data symbol in the frequency domain by transmitting all the chips of a spread symbol at the same time into a large number of narrow subchannels. This way, relatively high data rate can be supported by increasing the duration of each symbol, which drastically reduces inter-symbol interference.

3.2 Recent work

The longest running underwater acoustic networking experiments have been conducted as part of the Seaweb project [29, 30]. This series of experiments used FDMA in the beginning due to modem limitations but the limited bandwidth availibility and frequency-selectivity of the underwater acoustic channel made this undesirable. Recent Seaweb experiments use a hybrid form of TDMA-CDMA along with MACA type handshakes. The Seaweb deployment is the most extensive and includes not only a MAC-layer but also has neighbor discovery schemes for constructing dynamic routing tables using a centralized server architecture [30]. Seaweb is capable of operating over a period of several days and in regions that are in excess of 100 km^2.

Rapidly deployable single-hop star-topology AUV networks are described in [31]. Once deployed these networks operate over a range of approximately 5km^2; a gateway bouy provides operator control for the AUVs using TDMA for low-rate commands and high-rate for data communication. ACMENet [32] also uses a centralized TDMA protocol with adaptive data rates and power control.

A Slotted FAMA technique proposed by [33] works by adding time slots to FAMA to limit the impact of propagation delays encountered in the underwater channel. Another proposed approach [34] is to limit the impact of long RTS/CTS

CHAPTER 3. MAC PROTOCOLS

handshake packets by making handshake timing proportional to the separation of the communicating nodes.

Another potential approach is using combined TDMA-CDMA clusters, as was done in the Seaweb experiments. This allows shortening the TDMA slot lengths but increases overhead and the potential for interference from a neighboring cluster.

3.3 Future directions

The limited bandwidth and high propagation delays in underwater acoustic channels raise the need for cross-layer optimizations and adaptive parameter settings. Control packets in MAC protocols can be used as a means to sample the channel and setup the network parameters based on them by measuring propagation delays to set timeouts, received signal strength to set transmit power and signal-to-noise ratios to setup coding rates. Networks like Seaweb and ACMENet already include some provisions for adaptation and can serve as a model to develop adaptive protocols further.

Frequency-dependence of attenuation in the underwater channel [8] presents some advantages that could be exploited as well. A dual-frequency modem could be utilized with a lower-frequency transducer used for long-range communications and a high-frequency transducer for short-range high-bandwidth links. This could lead to not only power efficiency gains but also an increased throughput. Some new approaches also try to preserve the broadcast nature of the channel by using TDMA to share control and data for collective behavior of AUVs in an underwater long-wave radio network [35].

Chapter 4

Network Topologies, Mobility and Sparsity

Terrestrial networks generally assume fairly dense, continuously connected coverage of an area using inexpensive, stationary nodes. However, the costs associated with deployment and maintenance of underwater acoustic networks result in most underwater networks having sparse deployments. Furthermore, even static underwater networks have to deal with natural ocean currents that bring in an added degree of complexity that is generally attributed only to mobile deployments.

Large areas of interest, in case of oceanic surveys, and high cost of ship-based surveys has also led to the widespread use of mobile AUVs that need not only access to data channels but also methods for periodic localization signals to be made available for accurate navigational purposes. Due to limits of the physical channel, navigation and communication signals often share frequency bands in underwater acoustic networks and this combined demand on the channel further limits on the density of nodes in a network.

The sparsity and mobility of underwater acoustic networks gives rise to disruption-tolerant networks (DTNs); though a recent field of survey, DTNs are becoming increasingly studied by the WSN community. The DTN area may also provide insight which could be useful in design and operation of underwater acoustic networks. For example, it is widely known from study of DTNs that mobility patterns influence performance of a network. Finally, the sparsity and mobility also implies the necessity of a new operating regime for MAC protocols since it may be required in some scenarios to prioritize access for AUVs that are within communication range only briefly, to maintain long-term fair access to the channel.

This chapter looks at the different topologies that are commonly used by static underwater acoustic networks and also the knowledge made available by DTNs, as applicable in underwater acoustic networks. It presents some of the latest issues encountered in the sharing of localization and data signals within the same channel.

4.1 Static Networks

The network topology is in general a crucial factor in determining the energy consumption, the capacity and the reliability of a network. Hence, the network topology should be carefully engineered and post-deployment topology optimization should be performed, when possible.

Underwater monitoring missions can be extremely expensive due to the high cost of underwater devices. Hence, it is important that the deployed network be highly reliable, so as to avoid failure of monitoring missions due to failure of single or multiple devices. For example, it is crucial to avoid designing the network topology with single points of failure that could compromise the overall functioning of the network.

The network capacity is also influenced by the network topology. Since the capacity of the underwater channel is severely limited, it is very important to organize the network topology such a way that no communication bottleneck is introduced.

4.1.1 2-D Underwater Sensor Networks

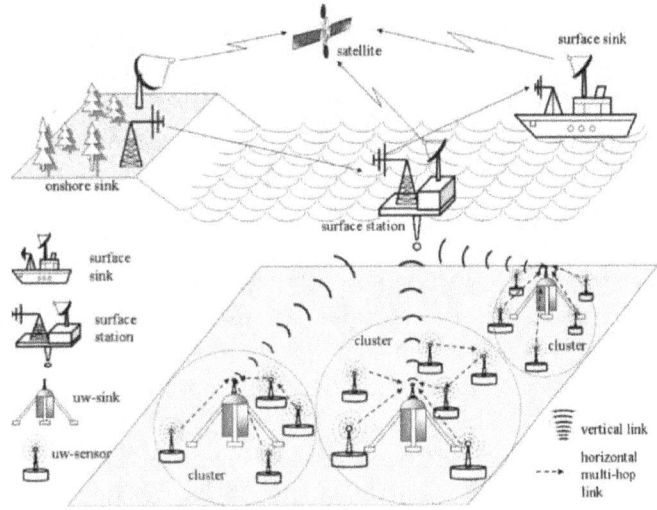

Figure 4.1: A Typical 2-D Underwater Network [1]

In most 2-D underwater sensor networks, for example Figure 4.1, a group of sensor nodes are anchored to the bottom of the ocean with deep ocean anchors and these are interconnected to one or more underwater sinks (uw-sinks) using acoustic links.

Uw-sinks are in charge of relaying data from the ocean bottom network to a surface station from which the data may be easily accessed.

In order to provide both surface and ocean-bottom communications, uw-sinks are equipped with a vertical and a horizontal acoustic transceiver. The horizontal transceiver is used to communicate with the sensor nodes and the vertical link is used to relay collected data to a surface station. The surface station is equipped with an acoustic transceiver capable of handling multiple parallel communications with the deployed uw-sinks in the network. These surface stations can be equipped with long range RF and/or satellite transmitters to communicate with an onshore or ship-based sink.

Sensors can be connected to uw-sinks via direct links or through multi-hop paths in case the transmission distance is too large. Each sensor sends the gathered data to the selected uw-sink, either directly or by relaying through intermediate nodes. Direct links are normally not preferred in order to introduce power efficiency within the network and also because direct links are very likely to reduce the network throughput as a result of the increased acoustic interference due to high transmission powers that would be needed in case of long transmission distances. Every network device usually takes part in a collaborative process whose objective is to diffuse topology information such that efficient and loop free routing decisions can be made at each intermediate node [2].

4.1.2 3-D Underwater Sensor Networks

Two-dimensional networks suffer from the shortcoming that they are unable to observe phenomena that does not occur at the ocean bottom. Three-dimensional underwater networks are deployed to overcome this shortcoming. In three-dimensional underwater networks, sensor nodes float at different depths in order to observe a given phenomenon. The depth of the nodes can be regulated by attaching them to surface bouys and then modifying the weight of the node to regulate the depth. This solution allows rapid deployment of the network but multiple floating buoys can be an obstruction in busy shipping lanes and floating buoys are vulnerable to weather and can also move due to ocean currents.

Annother approach is to anchor the sensors to the ocean bottom and equip it with a floating buoy that can be inflated by a pump to regulate the depth; such a topology is presented in Figure 4.2. The depth of the sensor can then be regulated by adjusting the length of the wire that connects the sensor to the anchor, by means of an electronically controlled engine that resides on the sensor [2].

CHAPTER 4. NETWORK TOPOLOGIES, MOBILITY AND SPARSITY

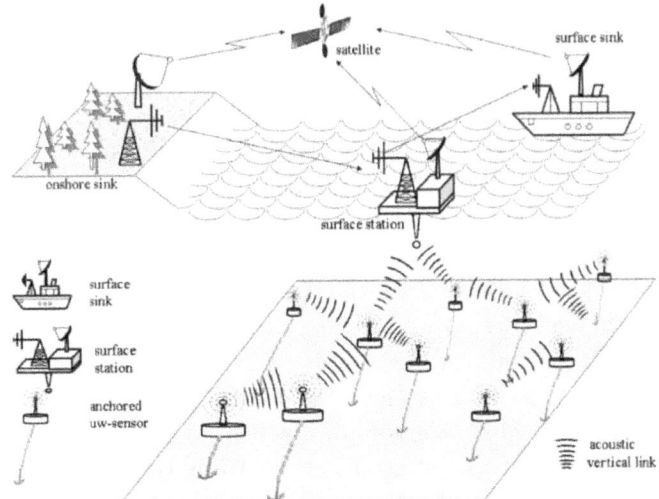

Figure 4.2: A Typical 3-D Underwater Network [2]

The following challenges need to be overcome in order for 3D coverage and network efficiency to be maximized:

- Sensing coverage. Sensors should collaboratively regulate their depth in order to achieve 3D coverage of the ocean column, according to their sensing ranges. Hence, it must be possible to obtain sampling of the desired phenomenon at all depths.

- Communication coverage. Since in 3D underwater networks there may be no notion of an uw-sink, sensors should be able to relay information to the surface station via multi-hop paths. Thus, network devices should coordinate their depths in such a way that the network topology is always connected, i.e., at least one path from every sensor to the surface station always exists.

The diameter, minimum and maximum degree of the reachability graph that describes the network can be derived as a function of the communication range, while different degrees of coverage for the 3D environment can be characterized as a function of the sensing range.

4.2 Disruption-Tolerant Networks

Most underwater networks comprise of mobile and sparse deployments [1, 2] and as a result DTNs arise since the link-layer coverage becomes partitioned. When two nodes are in communication range of each other, they have transfer opportunities from the time they discover one another until they are out of acoustic range. Even though radio networks are effected, in case of underwater acoustic networks the amount of data that can be transferred during each opportunity is especially the most constrained resource due to the limited bandwidth availibility in the channel. In order to ensure data delivery, a series of dynamic or pre-arranged meetings between nodes can form a path to a destination. If meetings are frequent and common, then the total throughput that can be delivered by the network can be reasonable for data that remains valuable after long delays. DTNs can also be used to connect geographically remote clusters of nodes.

Even though DTNs have primarily been researched under the assumptions of radio-based terrestrial networks, many of the techniques are directly applicable to underwater networking as well. Most approaches replicate packets epidemically during intermittent opportunities for transfer but at the same time, most of the protocols attempt to limit replication to only the nodes that appear to have some path to the destination. Most approaches to discovering paths to destination nodes make use of historic information regarding the past meetings of nodes. Some concepts from DTNs that can be applied in underwater networking include those such as, removing old packets representing delivered data from the network using broadcast acknowledgments and using network coding to efficiently take advantage of multiple paths [6].

The performance of a DTN can be greatly improved by making use of mobile nodes that have controllable movements. A system deployed on an AUV, discussed in [36], in a test pool plans, a route to visit stationary underwater nodes in known locations. Authors of [37, 38] investigate DTN routing based on ferries that operate on pre-planned paths designed to optimize network performance and known to all other nodes. A method for robotic agents to dynamically adjust movements according to perceived network conditions and according to multiple network objectives, such as maximizing delivery rate and minimizing delivery latency is proposed in [39, 40].

4.3 Data and Localization Signals

With the increased usage of AUVs as mobile nodes in underwater networks, it is essential to understand the dynamic between data signals produced by the network and the localization signals that are required by AUVs for navigational information. Since navigation information cannot be supplied by GPS underwater it is generally supplied by acoustic transponders in a long-baseline configuration. In typical applications [41] the vehicles normally ping navigation transponders about three times per minute to minimize navigation errors. Due to the frequency and range dependent attenuation of the channel, high-resolution navigation systems and high-throughput communications systems covering a region of a given size will generally use similar center frequencies, and hence often have interfering signals. MAC protocols in mobile underwater networks therefore need to be able to share the channel between network communications and navigation signals. When many vehicles are in an area, each vehicle must reduce the rate at which it pings localization transponders, which leads to navigation errors; methods need to be devised to overcome such shortcomings and underwater networks can be leveraged to further enhance localization information available to AUVs in such situations.

A passive localization and navigation method is described in [41] where a large number of vehicles passively share navigation signals in a manner similar to GPS without each vehicle actively pinging a transponder. In this method, when a vehicle needs more accurate location information, they can request a slot for an active long-baseline transponder ping. High-quality inertial navigation information from a master vehicle can be transmitted to companion vehicles, using synchronized hardware clocks and one-way travel-time measurements in order to aid multi-AUV cooperative missions [42].

A collaborative AUV mapping approach proposed in [43] makes AUVs share their individual maps over the broadcast network implemented in the acoustic channel, in the process making travel time measurements and creating a unified map, which can in turn be used for routing. The ICoN protocol outlined in [44] works by prioritizing navigation and communication packets to ensure that AUVs receive the necessary level of navigation information while ensuring that the network still remains responsive to command packets.

Chapter 5

Existing Evaluation Methodologies

All network protocols, topologies and methodologies need a robust evaluation methodology in order to test their performance and capabilities. Since the deployment costs associated with underwater networks is quite high, it is important for these test beds to provide accurate test results, to be rapidly deployable, allow for quick changes and modifications to the network and provide detailed in-depth analysis of the traffic, power consumption and other network parameters.

Though there is no perfect replacement for offshore testing of a network by actual deployment, the exhorbitant costs of offshore testing, maintenance and possible reconfigurations makes simulation environments an excellent tool to develop and test an underwater network before deployment. Due to the nascent nature of the underwater networking area, there are not many simulators available for the underwater acoustic channel but this chapter provides details on the few simulation tools available. Furthermore, to bridge the gap between offshore testing and simulation results low-cost laboratory test beds are also useful and the chapter provides some insight in currently usable laboratory test beds as well.

5.1 Simulation Environments

5.1.1 NS-2 Based Underwater Channel Simulator

The NS-2 simulator is a popular tool used for simulating complex networks and also wireless sensor networks. Authors of [7] present an implementation of an interface and channel model for underwater acoustic networks in the NS-2 network simulator.

As part of their work the authors construct a channel model that is based upon the Thorp equation [17, 18] for calculating the attenuation coefficient that effects all propagation parameters in the underwater acoustic channel. Since the underwater acoustic channel is quite different from the radio channel, which the NS-2 simulator is designed for, the authors design mathematical models that provide necessary information required by NS-2 for modelling the channel and physical layer.

To develop an accurate channel and physical model, the authors define a propagation model that calculates the speed of sound based upon the equations that were previous presented in Chapter 2. The propagation model is then used along with the previously mentioned Thorp model to determine the parameters such as transmission strength, signal-noise-ratio, attenuation and etc.

The simulator also models ambient noise realistically by taking into account the effect of external sources such as shipping, wind, thermal and turbulence noise. Using mathematical models that predict these parameters the NS-2 based simulator is able to accurately model the noise characteristics of the underwater acoustic channel.

The modulation provides bitrate and bit-error calculations that are needed by NS-2 to correctly simulate a network setup. Once the frequency dependent attenuation constant, ambient noise, propagation delay and transmit power are available, the Shannon theorem [8] is used by the simulator to calculate the bitrate used by NS-2.

The NS-2 based simulator currently provides support only for MAC and PHY layer implementations and is provided along with an implementation of FDMA and ALOHA protocols. There is no support for routing and transport layer protocols available and a protocol stack needs to be implemented as well. Being based on the NS-2 simulator there is full support for testing network performance, including collisions and interference. Upon completion of a simulation the simulator provides a NS-2 trace file that can be analyzed in detail to test and evaluate network performance and shortfalls. This simulator provides an excellent basis for building further test beds that more accurately model the underwater acoustic channel.

5.1.2 OPNET Based Underwater Channel Simulator

An underwater acoustic local area network is designed and tested using OPNET's Radio Modeler in [45]. The authors of this paper design a network that consists of master and sensor nodes which utilize battery powered modems and rely upon the model of the Datasonics ATM-875 modems within the simulation.

For the purpose of the simulation, it is assumed that the network nodes are stationary and that the channel is slowly varying and stays constant during a packet interval. Similar to the NS-2 simulator, the authors design and implement their own path loss, background noise and propagation delay model using the Radio Pipeline stages of the OPNET simulator. The Thorp equation is used in order to model the path loss that occurs during transmission and the background noise is assumed to be constant during the length of the simulation. The speed of sound in water is taken as a constant velocity of 1500 m/s for the purpose of the simulation.

Even though this OPNET based simulation provides a good basic platform for simulating the underwater acoustic channel, due to its limitations of depending upon the Thorp equation, which does not take into account the complex dynamics which effect the propagation loss, and also the inaccurate method of modeling noise and sound velocity as a constant, the simulator is not robust enough to provide dependable results that may be reproduced accurately in off-shore testing.

5.1.3 MATLAB Based Underwater Channel Simulator

MATLAB based simulations of the underwater acoustic channel are quite popular in literature, however, mostly these are highly application specific and deal with simulating the lower layers only. A more general purpose underwater acoustic channel simulation environment based on MATLAB that incorporates multipath propagation, surface and bottom reflection coefficients, attenuation, spreading and scattering losses as well as the transmitter/receiver device employing Quadrature Phase-Shift Keying modulation techniques is presented in [46].

Even though this simulation environment provides quite an in-depth simulation of the communication channel, it does not provide a method for defining custom topologies, power models or methods for monitoring other factors like packet transmissions, losses and collisions that might interest the networking community and might even impact the performance of a network in the underwater channel. Additionally, no support for any routing protocols is made available in the simulation environment either. Since AUVs are expected to be one of the largest users of underwater acoustic networks the issue of mobility is also a very important one to be investigated and the ability to simulate node mobility is also absent from within this simulator.

Furthermore, just as any other simulation based in MATLAB, this environment also suffers from slow processing times. All these factors highlight the need for a more efficient simulation environment.

5.1.4 NetMarSys - Networked Marine Systems Simulator

The NetMarSys [47] simulator is designed and used by the Institute for Systems and Robotics in Portugal. The simulator is a software suite intended to simulate different types of cooperative missions involving a variable number of heterogeneous marine craft, each with its own dynamics. The high level of detail to which the environment can be modeled allows to take into account both the effect of water currents on the vehicle dynamics as well as the delays and environmental noise that affect underwater communications.

Though NetMarSys provides an excellent platform for defining models for mobile nodes, it lacks the necessary sophistication to accurately simulate the effects that ocean dynamics have on the underwater acoustic communication channel. Furthermore, the simulator uses an over simplified model for calculating the propagation delay by using the following equation:

$$\tau = \frac{d}{c}$$

where τ is the delay, d is the distance between nodes and c is the speed of sound in water, which is used as a constant of 1500 m/s in the simulator.

This shortcoming of oversimplifying the underwater acoustic channel model makes the results provided by the simulator not an accurate profile. However, the mobility features of the simulator are something that would be extremely useful to be included within other simulators as well.

5.2 Laboratory Test-beds

Underwater acoustic networks operate in rapidly changing and hostile environmental conditions, furthermore, the nodes are expensive to manufacture, deploy, maintain and retrieve [6]. These reasons coupled with the need to review and possibly even redesign some aspects of a network during it's inception state, present the need for a robust laboratory test bed that would provide accurate test results on the network's performance and would also allow for rapid prototyping of new ideas, topologies and protocols while maintaining an accurate model of the underwater acoustic channel.

5.2.1 Aqua-Lab

Aqua-Lab [48] is an underwater acoustic sensor network lab testbed designed and hosted at the UnderWater Sensor Network Lab at the University of Connecticut. At an overview level, Aqua-Lab consists of a water tank, acoustic communication hardware and software that controls the configuration and operation of the testbed. As part of the software environment the Aqua-Lab consists of an emulator that provides programming interfaces and emulates realistic underwater network settings.

Acoustic modems and transducers form the communication hardware, the operation of which is encapsulated by software APIs that provide an abstract layer for users so that custom applications could be developed without knowing the exact mechanisms of the underlying acoustic physical layer. The emulator is capable of emulating different network topologies, propagation delay, and attenuation. The Aqua-Lab is based on the WHOI Micro-Modem acoustic modems. A C library provides an interface to the acoustic modems and the operations to set up options such as the frequency band, baud rate, data request timeout, sleep-mode operation, opening a port for communication, closing a port, pinging other modems, reading messages, and writing messages.

CHAPTER 5. EXISTING EVALUATION METHODOLOGIES 32

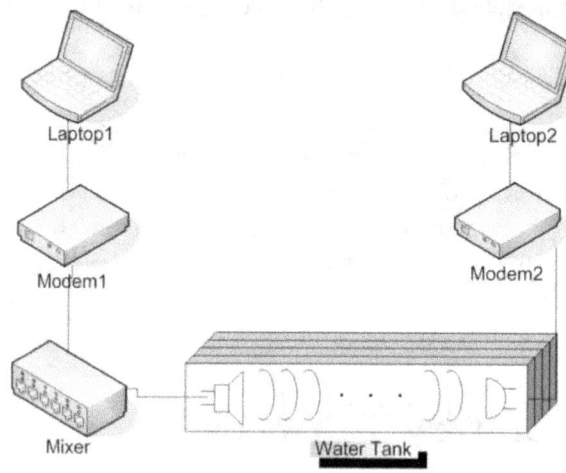

Figure 5.1: Aqua-Lab Testbed Setup [48]

The hardware setup for the testbed consists of the following:

- WHOI Micro-Modem - allows for acoustic communication between nodes in the Aqua-Lab using either a high-data-rate or low-data-rate mode.

- Underwater speaker - with a frequency range from 20Hz to 32KHz.

- Hydrophone - supporting frequencies from 20Hz to 100KHz.

- Sound mixer - is utilized to emulate different underwater environments and multiplexing multiple signals.

- Aquarium - of size $2m^3$ in size holds approximately two tons of water.

- Server - to control the acoustic modems and to execute the emulator to setup complex network scenarios.

Figure 5.1 provides a logical overview of the setup for the Aqua-Lab testbed. The results presented by the authors confirm that the test-bed provides similar results to that observed in offshore testing, thereby making Aqua-Lab a good model to follow for designing a test-bed for controlled laboratory tests intended for the underwater acoustic channel.

Part III

Underwater Acoustic Channel Model Development, Analysis and Simulation

Chapter 6

Model Development and Numerical Analysis

An accurate understanding and modeling of the underwater acoustic channel is the basis upon which all work for underwater networks is based. There exist several models for calculating and predicting the attenuation, which effects all other aspects of the underwater acoustic channel model. Furthermore, parameters from frequency, distance, depth, acidity to salinity and temperature of the underwater environment effect how the channel acts and in turn also result in changing network performance. It is as such important to understand the relationship between all these parameters and the effect they have on the performance of a network that uses the underwater acoustic channel.

As such, as a basis for further work, it is necessary to analyse the different channel models available and compare their results with each other in a numerical form in order to obtain an understanding of which channel models are the most appropriately suited for predicting the performance of an underwater channel. This chapter formulates the different underwater channel models, numerically compares them and then arrives to a conclusion based on the observed results as to which models are the most suitable for usage.

6.1 The Underwater Acoustic Propagation Model

The performance predicted by an underwater acoustic channel model is greatly dependent upon the propagation model that is chosen. The greatest changes in the acoustic models are caused by the the attenuation model that is chosen. In this section the basic underwater acoustic propagation model based upon the attenuation models that are discussed in details within Chapter 2 is formulated. The propagation model formulated in this section forms the basis for the overall channel model that is utilized to characterize the underwater acoustic channel.

6.1.1 Propagation Delay

For most purposes the speed of sound in water is taken to be approximately 1500 m/s. While this is accurate within a certain range, the underwater channel is an extremely complex environment that is effected by many varying factors, primarily temperature, salinity and depth [11, 9] and furthermore each of these factors may also be interdependent or varying across the ocean. It is, as such, important to have an accurate model of the effects of these parameters on the speed of sound in water.

Since the MacKenzie equation discussed in Chapter 2 provides an estimate of the speed of sound in water with an error in the range of approximately 0.070 m/s, it has been chosen as the basis of all propagation delay modeling for this investigation. Using the MacKenzie equation, obtained from Equation 2.1, the propagation delay that can be observed in an underwater acoustic channel can be easily obtained, if the thermocline and halocline are also defined.

6.1.2 Propagation Loss

	Spherical	Cylindrical	Practical
k	2	1	1.5

Table 6.1: Values for representing types of geometrical spreading via the geometrical spreading coefficient k

The transmitted acoustic signal between sensor nodes in a network reduces in overall signal strength over a distance due to many factors like absorption caused by magnesium sulphate and boric acid, particle motion and geometrical spreading. Propagation loss is composed majorly of three aspects, namely, geometrical spreading, attenuation and the anomaly of propagation. The latter is nearly impossible to model and as such the attenuation, in dB, that occurs over a transmission range l for a signal frequency f can be obtained by modifying Equation 2.14 to represent also the geometrical spreading that occurs over a particular range:

$$10 \log A(l, f) = k \cdot 10 \log l + l \cdot 10 \log \alpha \qquad (6.1)$$

where α is the absorption coefficient in dB/km, which can be obtained from models specifically characterizing it, and k represents the geometrical spreading factor. This geometrical spreading factor can be substituted with values shown in Table 6.1 in order to represent accurately the type of spreading that occurs. The overall propagation loss can be easily obtained when Equation 6.1 is used along with an appropriate attenuation model that provides the absorption coefficient α.

6.1.3 Absorption Coefficient

Attenuation by absorption occurs due to the conversion of acoustic energy within sea-water into heat. This process of attenuation of absorption is frequency dependent since at higher frequencies more energy is absorbed. There are several equations describing the processes of acoustic absorption in seawater which have laid the foundation for current knowledge. Each of these equations has over time improved the applicability and accuracy of mathematically predicting the absorption of sound in sea water. Each mathematical model obtains the signal absorption coefficient according to environmental and signal characteristics. In this section our propagation model based upon the attenuation by absorption models discussed in Chapter 2 is formulated.

6.1.3.1 Thorp Model

In order to obtain the absorption coefficient in dB/km from the Thorp model provided in Equation 2.7 and also have it directly applicable in the propagation loss model of Equation 6.1, the original equation is modified to take the form:

$$10 \log \alpha = \frac{0.1 f^2}{1 + f^2} + \frac{40 f^2}{4100 + f^2} + 2.75 \times 10^{-4} \cdot f^2 + 0.003 \qquad (6.2)$$

This model is very simple to implement and only utilizes a dependence upon the signal frequency. It is designed to be most accurate for a temperature of 4°C and a depth of approximately 1000m.

6.1.3.2 Fisher & Simmons Model

As with the Thorp model, the original Fisher & Simmons model expressed in Equation 2.8 is also modified to take the form:

$$10 \log \alpha = A_1 P_1 \frac{f_1 f^2}{f_1^2 + f^2} + A_2 P_2 \frac{f_2 f^2}{f_2^2 + f^2} + A_3 P_3 f^2 \qquad (6.3)$$

The model expressed in Equation 6.3 provides the absorption coefficient in dB/km. The additional coefficients in Equation 6.3, A_1, A_2, A_3, P_1, P_2, P_3, f_1, f_2 can be obtained from Section 2.2.2.3 of Chapter 2.

6.1.3.3 Ainslie & McColm Model

The Ainslie & McColm model provided by Equation 2.9 is modified to take the following form in order to provide a result in db/Km:

$$\begin{aligned} 10 \log \alpha = & \; 0.106 \frac{f_1 f^2}{f_1^2 + f^2} e^{\frac{pH-8}{0.56}} \\ & + 0.52 \left(1 + \frac{T}{43}\right) \left(\frac{S}{35}\right) \frac{f_2 f^2}{f_2^2 + f^2} e^{\frac{-D}{6}} \end{aligned} \qquad (6.4)$$

$$+4.9 \times 10^{-4} f^2 e^{-\left(\frac{T}{27}+\frac{D}{17}\right)}$$

The coefficients for the above equation may be obtained from Section 2.2.2.4 of Chapter 2.

6.1.4 Ambient Noise Model

Ambient noise in the ocean can be described as Gaussian and having a continuous power spectral density (p.s.d.). The four most prominent sources for ambient noise are the turbulence, shipping, wind driven waves and thermal noise. The p.s.d. in dB re μPa per Hz for each of these is given by the formulae [49] shown below:

$$10 \log N_t(f) = 17 - 30 \log f \tag{6.5}$$

$$10 \log N_s(f) = 40 + 20(s - 0.5) + 26 \log f - 60 \log(f + 0.03) \tag{6.6}$$

$$10 \log N_w(f) = 50 + 7.5 w^{\frac{1}{2}} + 20 \log f - 40 \log(f + 0.4) \tag{6.7}$$

$$10 \log N_{th}(f) = -15 + 20 \log f \tag{6.8}$$

The ambient noise in the ocean is colored and hence different factors have pronounced effects in specific frequency ranges. In the noise model equations utilized for this study the colored effect of noise is represented by N_t as the turbulence noise, N_s as the shipping noise (with s as the shipping factor which lies between 0 and 1), N_w as the wind driven wave noise (with w as the wind speed in m/s) and N_{th} as the thermal noise.

Turbulence noise influences only the very low frequency region, $f < 10$ Hz. Noise caused by distant shipping is dominant in the frequency region 10 Hz -100 Hz. Surface motion, caused by wind-driven waves is the major factor contributing to the noise in the frequency region 100 Hz - 100 kHz (which is the operating region used by the majority of acoustic systems). Finally, thermal noise becomes dominant for $f > 100$ kHz.

The overall noise p.s.d. may be obtained in μPa from:

$$N(f) = N_t(f) + N_s(f) + N_w(f) + N_{th}(f) \tag{6.9}$$

The noise p.s.d. may be used along with the signal attenuation to arrive at values that characterize the channel performance. The obtained value may be converted to dB by following the method described in Section B.2 of Appendix B.

6.2 The Underwater Acoustic Channel Model

Since the underwater acoustic channel is locally time varying, there exists no single character for the channel that could be globally used as a model. This makes it important to characterize the underwater acoustic communication channel in order to determine the effects of local environmental phenomenon on achievable performance. This performance of the channel can be characterized by properties that include received signal power (which is dependent on the transmission power), signal-to-noise ratio (SNR) and the capacity bound.

6.2.1 Received Signal Power

The path loss represented by Equation 6.1 is the attenuation that occurs on a single unobstructed propagation path. As such, if a signal with frequency f is transmitted over distance l with a power P_{tx} then we can calculate the arriving signal power P_{rx} in dB as:

$$10 \log P_{rx} = 10 \log P_{tx} - 10 \log A(l, f) \tag{6.10}$$

The result obtained from Equation 6.10 takes into account only the case for a directional transmission, i.e., the most direct propagation path from transmitter to receiver. However, in case of a transmission that is not directional needs to be modelled, this equation can be extended for the indirect routes as well. At present, in this work the focus is only upon the directional transmission model in order to obtain the received signal's p.s.d.

Since the received signal power is dependent upon the propagation loss factor, the attenuation model choice also adds a dependence upon depth, temperature, salinity and acidity of the specific oceanic region that is of interest.

6.2.2 Signal-to-noise ratio

Using knowledge of the signal attenuation $A(l, f)$ and the noise p.s.d. $N(f)$ the SNR observed at the receiver may be calculated. Extending Equation 6.10 we can arrive at the following relationship for obtaining the SNR in dB:

$$10 \log SNR(l, f) = 10 \log P_{tx} - 10 \log A(l, f) - 10 \log N(f) \tag{6.11}$$

where $SNR(l, f)$ is the SNR over a distance l and transmission center frequency f. Similar to the received signal power, the attenuation model choice also adds a dependence upon depth, temperature, salinity and acidity of the specific oceanic region that is of interest, for the SNR.

6.2.3 Optimal Transmission Frequencies

The attenuation noise (AN) factor, given by $-[10 \log A(l, f) + 10 \log N(f)]$ from Equation 6.11, provides the frequency dependent part of the SNR. By close analysis of this relationship, it can also be determined that for each transmission distance l there exists an optimal frequency at which the maximal narrow-band SNR is obtained. Since the SNR is inversely proportional to the AN factor, the optimal frequency is that for which the value of $1/AN$ (represented in dB re μPa per Hz) is the highest over the combination of a certain distance, $f_o(l)$. Using these optimal

CHAPTER 6. MODEL DEVELOPMENT AND NUMERICAL ANALYSIS

frequencies one may choose a transmission bandwidth around $f_o(l)$ and adjust the transmission power to meet requirements of a desired SNR level.

All the formulation in this analysis work is based upon the optimal frequencies $f_o(l)$, however, it may be extended to any desired frequency by replacing $f_o(l)$ with the chosen transmission frequency $f_{tx}(l)$ for a particular application.

6.2.4 Bandwidth

Authors of [8] present capacity as a 3 dB band heuristic definition in their work, and we utilize the same definition for calculating the channel capacity. As such, the available bandwidth is a range of frequencies around $f_o(l)$, such that the difference of $A(l, f_o(l))N(f_o(l))$ and $A(l, f)N(f)$ is within the bandwidth definition. Here we can define $f_{min}(l)$ as the smallest frequency for which $AN_{f_o(l)} - AN_f \leq 3$ holds true and $f_{max}(l)$ as the largest frequency f for which $AN_{f_o(l)} - AN_f \leq 3$ holds true as well. Thus, the transmission bandwidth $B(l)$, over a distance l, becomes:

$$B(l) = f_{max}(l) - f_{min}(l) \tag{6.12}$$

6.2.5 Channel Capacity

Usable channel capacity is undoubtedly one of the best metrics since it governs many aspects of network design and can lead to significant changes in topologies, protocols and access schemes utilized in order to maximize the overall throughput. As per the Shannon theorem the channel capacity C, i.e. the theoretical upper bound on data that can be sent with a signal power of S subject to additive white Gaussian noise is:

$$C = B \log_2\left(1 + \frac{S}{N}\right) \tag{6.13}$$

where B is the channel bandwidth in Hz and $\frac{S}{N}$ represent the SNR. The basic Shannon relationship shown in Equation 6.13 can be extended to be applicable in cases where the noise is dependent on frequency to take the form of:

$$C = \int_B \log_2\left(1 + \frac{S(f)}{N(f)}\right) df \tag{6.14}$$

If we assume a time-invariant channel for a certain interval of time along with Gaussian noise then we can obtain the total capacity by dividing the total bandwidth into multiple narrow sub-bands and summing their individual capacities. In this case each sub-band has a width of a small Δf which is centered around the transmission frequency and this can be obtained from the relationship defined in Equation 6.12.

Extrapolating from the above discussed Equations 6.12 and 6.14, we may now obtain the channel capacity over distance l from:

$$C(l) = \int_B \log_2\left(1 + \frac{P_{tx}}{A(l,f)N(f)B(l)}\right) df \tag{6.15}$$

6.3 Numerical Evaluation

A deeper understanding of the propagation and channel models provided in the previous sections of this chapter is important in order to optimally choose the models for simulating underwater networks that utilize the acoustic channel. In this section we numerically evaluate the equations that are proposed in the previous sections and also compare the results obtained by using different models to arrive at various conclusions.

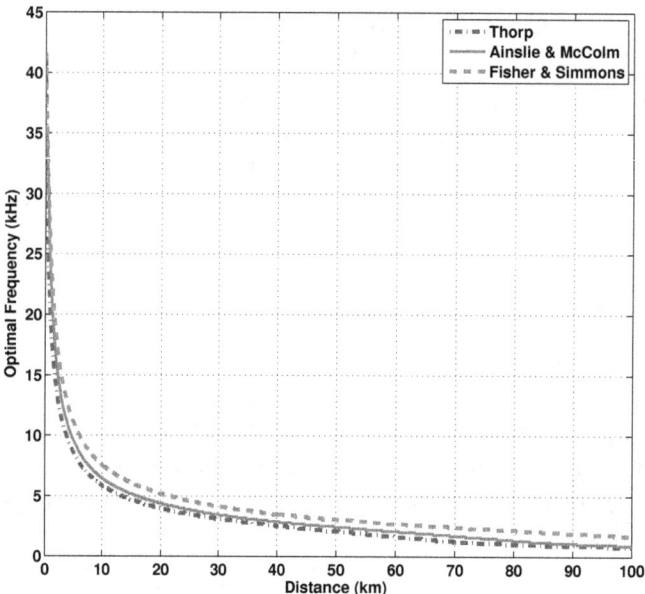

Figure 6.1: Optimal frequencies as predicted by the different channel models.

For the purpose of evaluation work, optimal frequencies for all transmission distances are utilized. Depth of 1 km, temperature of 4 °C, pH level of 8 and salinity of 35 ppt are utilized for all numerical evaluations, unless otherwise stated, in order to remain within the capability ranges of all the attenuation models.

6.3.1 Optimal Frequencies

The optimal transmission frequency provides the highest capacity and thereby most likely the best performance in the underwater acoustic channel. A comparison of the optimal frequencies predicted by each of the three propagation models is shown in Figure 6.1. An analysis of this plot reveals that the Fisher & Simmons model provides optimal frequencies which are higher than those predicted by the Ainslie & McColm and Thorp models. The results of the Ainslie &

McColm and Thorp models are very similar, however, it is already known from results obtained in Chapter 2 that the Thorp model does not provide results that are nearly as detailed as those provided by the Fisher & Simmons and Ainslie & McColm models since the Thorp model only takes into account the effects of transmission frequency on the attenuation coefficient, thereby limiting the predicted channel characteristics to effects of distance and frequency only.

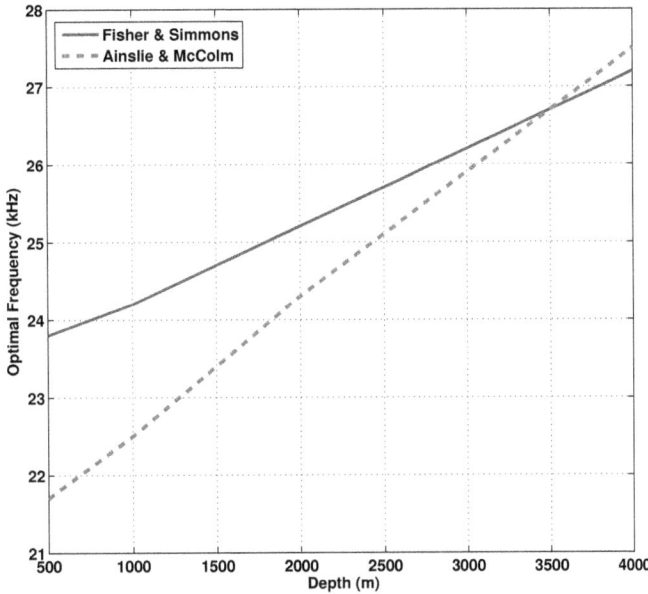

Figure 6.2: Optmial frequencies with changing depth.

The slight difference in the values between Ainslie & McColm and Thorp models can be attributed to the lack of channel characteristics parameters in the Thorp model. However, since we are mostly interested in the deep sea acoustic channel, it is important to take into account parameters such as depth, temperature, acidity and salinity in order to obtain an accurate representation of the underwater acoustic channel. This requirement makes the Ainslie & McColm and Fisher & Simmons models of greater interest to us. A comparison of the results provided by both these models shows a relatively large difference in the optimum frequency predicted. This can mostly be attributed to the fact that the Ainslie & McColm model is a simplified model, which obviously introduces some errors.

However, since the results provided by the Ainslie & McColm model are between those provided by the other two models, both of which are widely utilized in published literature, it appears to provide a good approximation of optimal frequencies (and thereby other channel results) as well. The additional

CHAPTER 6. MODEL DEVELOPMENT AND NUMERICAL ANALYSIS

computational overhead added by the multiple parameters, however, makes it more appealing to utilize the Thorp model when approximations are needed.

Using this conclusion as a basis, a comparison of the optimal frequencies as predicted by the Ainslie & McColm and Fisher & Simmons models with varying depth is provided in Figure 6.2. A transmission range of 500 m is used to obtain these results. It is clear from this plot that optimal frequencies increase with the increase in depth. Both models show this increasing trend, however, once again the Ainslie & McColm model predicts values lower than Fisher & Simmons for depths below 3500 m.

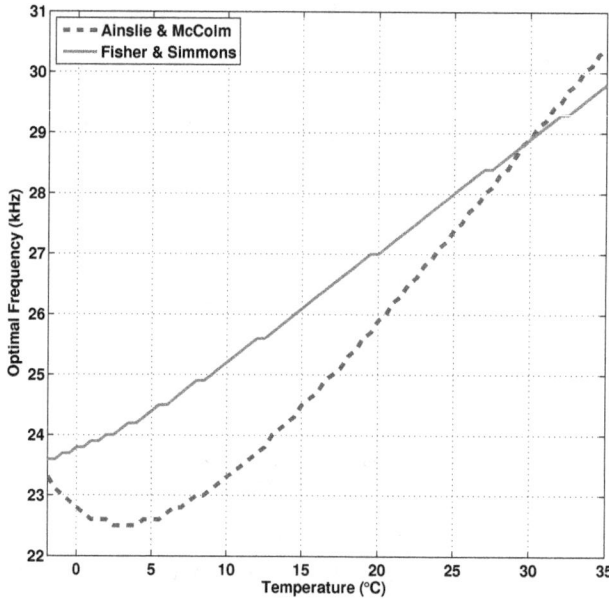

Figure 6.3: Optmial frequencies with changing ocean temperature.

Being interested in deep sea acoustic channels it is important to keep in context of analysis the halocline and thermocline as well. A detailed discussion of these two characteristics is provided in the Appendix, however, it is necessary to highlight that the ocean temperature varies between $-2°C$ and $36°C$ while the salinity only varies between 33-37 ppt with depth. The global oceanic acidity remains quite constant around a pH value of 8, besides specific regions such as underwater volcanoes. Keeping these values in mind a plot of optimal frequencies with varying temperature is provided in Figure 6.3. A transmission distance of 500 m is utilized to obtain results for this graph.

At first glance the results in Figure 6.3 make it appear as though the Ainslie & McColm model has an anomalous performance as compared to that of the Fisher & Simmons model since basic logic dictates that optimal frequency should

CHAPTER 6. MODEL DEVELOPMENT AND NUMERICAL ANALYSIS

increase with temperature as almost a linear relationship, much like that with depth. However, further analysis reveals that at 1 km depth the density of water is actually highest at 4°C, thereby representing a curve that looks almost parabolic curve [50]. Comparing this behavior of water density with the plot in Figure 6.3 reveals that the Ainslie & McColm provides results which adhere to this ideology. This result makes it clear without any doubt that the Ainslie & McColm model outperforms Fisher & Simmons, thereby making it the model of choice.

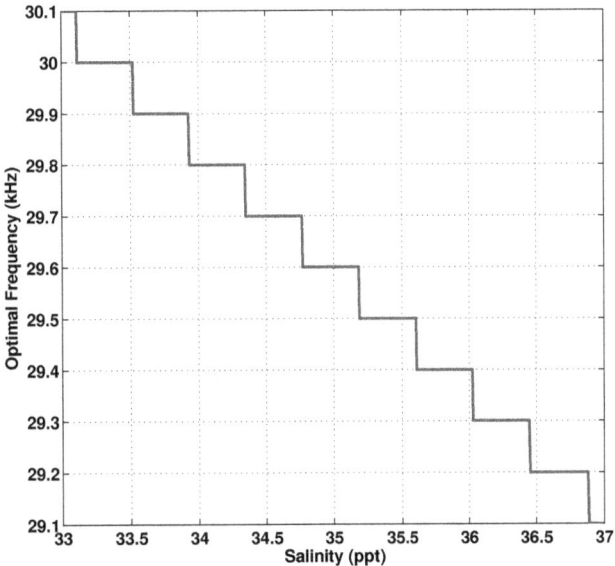

Figure 6.4: Optmial frequencies with changing ocean salinity.

The plot in Figure 6.4 clearly shows that the optimal transmission frequency reduces with changing salinity, however, the overall change of 1 kHz over the range of possible ocean salinities is quite neglegible and therefore does not pose a great effect on the performance of the underwater acoustic channel.

6.3.2 Bandwidth and Capacity

Optimal transmission frequencies were evaluated at various depths, temperatures and transmission distances in order to derive the patterns of effects that these parameters would have on the predicted channel capacity. For the purpose of evaluating ambient noise we assumed a shipping factor $s = 0.5$, to represent moderate shipping, and no wind-caused waves leading to a $w = 0$. Since we were able to determine in the previous section that the Ainslie & McColm model provides results that are more accurate than the Fisher & Simmons model, the Bandwidth and Capacity investigations were done using the Ainslie & McColm model. Furthermore, a comparison with the Thorp model was undertaken to provide a comparison between the models.

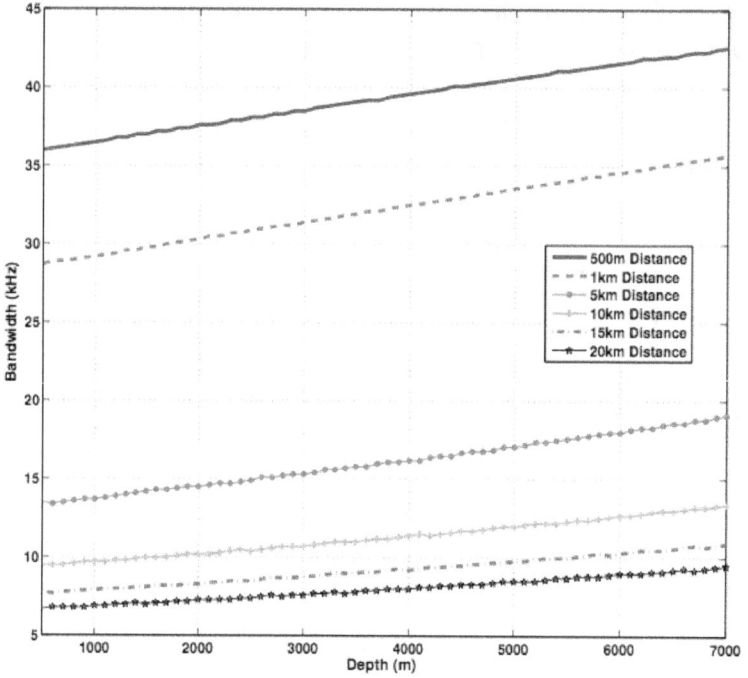

Figure 6.5: Effect of depth on available bandwidth.

In Figure 6.5 we plot the effects that depth and signal transmission distance have on bandwidth. The plot indicates that available bandwidth increases almost linearly with increasing depth, however, this linear effect is not constant over every transmission distance; the shorter the transmission distance, the more quickly bandwidth increases with depth. This clearly indicates that having network designs with short transmission distances would make better use of channel capacity and thereby provide an overall higher throughput. Depending upon the depth of the transmitting nodes the channel bandwidth can vary even as high as 11 kHz for a particular distance; on an average the bandwidth varies by 5 kHz as an effect of depth and transmission distance.

Since the Ainslie & McColm model also allows us to evaluate effects of temperature on bandwith and capacity we evaluated the variance of bandwidth with temperature and multiple transmission distances. Figure 6.6 shows a plot that displays the effects of temperature on bandwidth when multiple signal transmission distances are used. For the purpose of this evaluation a fixed depth of 500 m and temperatures ranging between -2°C and 36°C were used.

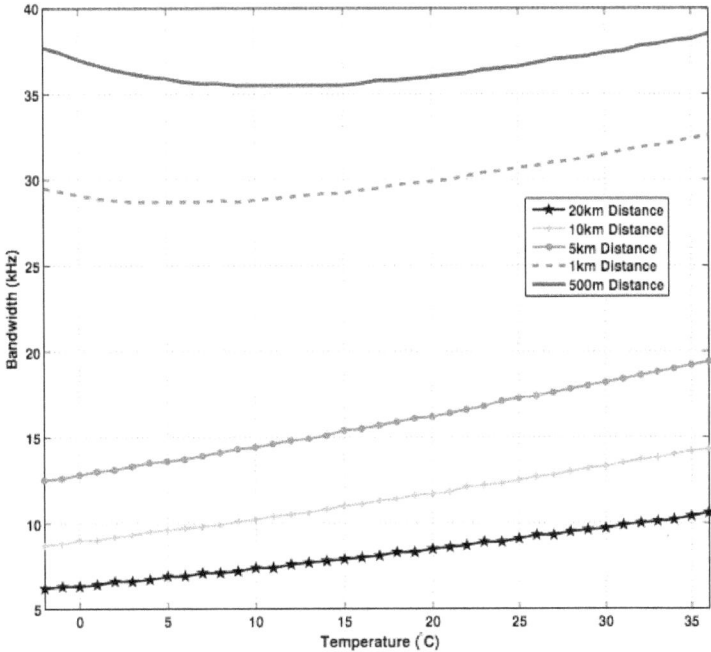

Figure 6.6: Effect of changing temperature on bandwidth.

From the plot in Figure 6.6 it becomes quite clear that increasing temperature generally causes bandwidth to increase, however, this relationship is not linear. Furthermore, at comparatively shallow depths (at least till 1 km depth) the bandwidth decreases from -2°C till a certain point before increasing again. Such non-constant increase or decrease in capacity makes it critical for underwater acoustic systems, at least mobile ones, to be designed with this in mind. Since the shallow water region normally comprises of the thermocline where temperature fluctuations can occur over the course of the day, shallow water networks would greatly suffer in performance unless they stick to the minimum bandwidth or follow adaptive bandwidth schemes.

The effects of temperature and depth on channel capacity are also investigated in order to have a more concrete picture. The channel capacity calculated using Equation 6.15 is plotted in Figure 6.7 for a comparison on the effects of depth and temperature. The plot presents the effects of temperature on capacity while also varying the depth but keeping the transmission distance fixed at 5 km. Just as with bandwidth, channel capacity increases with increasing depth and temperature also appears to have the same effect. In case of deep sea nodes or even mobile nodes, the network and protocols could be designed to take benefit of this fact by allowing deeper nodes to communicate at higher bandwidths to achieve an overall higher throughput.

6.3.3 Discussion

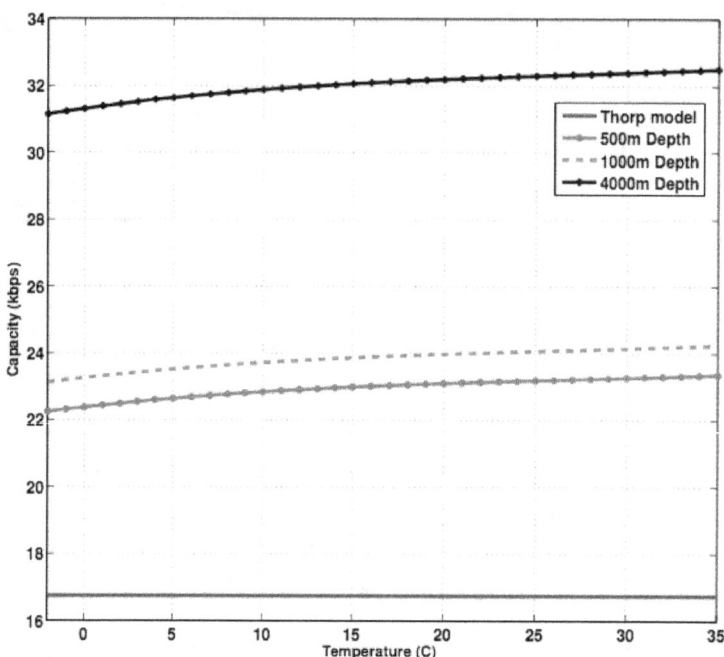

Figure 6.7: Effect of changing temperature on capacity.

An initial comparitive numerical analysis of the three different underwater acoustic propagation and channel models reveals that even though the Fisher & Simmons model is one of the most popularly utilized models in published literature, the results it provides are not as accurate to those provided by the Ainslie & McColm model. Furthermore, the ability to simulate effects of depth, temperature, salinity and acidity also make the Ainslie & McColm model comparatively highly desired. The Thorp model provides a good approximation of performance since it's results are closest to that of the Ainslie & McColm model at it's default channel characterisitc parameters. Since changing the model used changes the results significantly it is important to make an appropriate choice.

Furthermore, it is quite clear after the numerical evaluations that acidity does not have a pronounced effect on the channel performance characteristics, however, salinity, temperature and depth all have an effect that make it important to sample these values when designing a network topology or choosing acoustic modems for a particular application. It can also be said with high confidence that bandwidth and capacity decrease over longer transmission distances, while increasing depth provides higher bandwidths and capacity; the relationship between temperature and capacity and bandwidth is not linear but can be generalized to be mostly increasing with increasing temperatures.

The optimal transmission frequencies for longer transmission distances are lower but increasing the depth and termperature mostly increases the optimal transmission frequencies. The only anomaly in this relationship is due to the density of water being effected by temperature, thereby making the relationship between temperature, water density and optimal frequency a non-linear one.

Chapter 7

Software Implementation

Numerical analysis provides results which are interesting and can lead towards initial network design choices that could enhance overall network performance. However, network performance is not only dependent upon the physical characteristics of the underwater acoustic channel. In order to provide an overall performance analysis it is important to also evaluate the network statistics which result from media access control schemes, routing protocols, modulation schemes and other networking layers. In order to do so, it is important to build a software infrastructure that takes into account a complete acoustic propagation and channel model and implements them such that it can provide details on achievable (or achieved) data rates, performance of routing protocols, delivery ratio of packets and other characteristics. Furthermore, even though numerical models can represent propagation and physical layer issues, they fail to incorporate protocol issues such as collisions and multiple-access interference.

The NS-2 simulator, which is a highly popular tool used for simulating network performance, provides an excellent basis to develop a software implementation for simulating the underwater acoustic channel. This chapter disucusses the implementation of the AquaTools NS-2 underwater simulation toolkit which incorporates various channel models constructed in Chapter 6 into NS-2 in order to ensure that all conditions effecting the performance of a network can be analyzed. The trace files provided by NS-2 could be further useful in research to tweak and maximize network performance. It also covers the Wireless Simulation Server that was designed for the USARSim high-fidelity real-time robotics simulator in order to provide a platform to quickly test and evaluate the performance of an underwater acoustic network on mobile robot nodes, while keeping in mind the requirement for active environmental sensing and movement responses that a robot would have and which NS-2 cannot account for.

7.1 The AquaTools NS-2 Underwater Simulation Toolkit

The NS-2 simulator divides the channel and physical layer functions and characteristics into four components, namely Propagation, Channel, Physical, and Modulation. Figure 7.1 depicts this division, highlighting the characteristics within each component. The propagation component contains most of the characteristics of the signal propagation through the medium (including attenuation) and of the ambient noise. In addition to distance-dependent attenuation, in underwater channels the signal fading is also affected by the orientation of the link. This feature is also modeled in the propagation component. The characteristics exported to other components of the NS-2 model include the calculation of the received signal strength and the interference range of a signal.

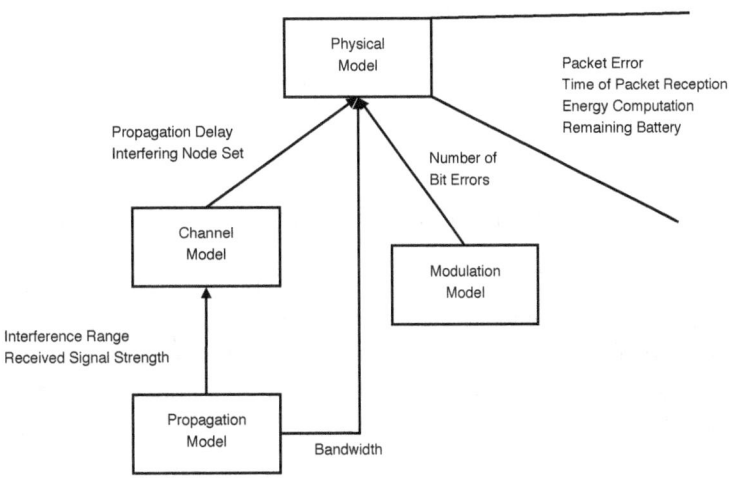

Figure 7.1: The NS-2 channel and physical layer functional model

The primary function of the channel model is to handle propagation delay calculations and to make use of the functions from the propagation model. The physical layer tracks energy consumption metrics and also calculates the transmission times. Unlike in radio models, where the bandwidth is assumed to be constant regardless of the transmitter-receiver distance, depth, ambient temperature, salinity or acidity and therefore no information for other layers is required, in an underwater network the link bandwidth does depend on the link length, and therefore bandwidth information from the propagation layer of NS-2 must be exposed to other components. Finally, the physical model calls the modulation model to calculate bit error probabilities given a received signal strength, modulation scheme, and level of noise. It is interesting to note that no standard modulation schemes are currently used in the majority of NS-2 simulations.

CHAPTER 7. SOFTWARE IMPLEMENTATION

The AquaTools toolkit uses dB re μPa as the unit of sound energy throughout the entire implementation since this a typical unit of signal strength which is used in acoustic communications. Accordingly, all quantities expressed here are in this unit, and all tunable parameters (for example, transmit power) are also given in dB re μPa as well.

7.1.1 Underwater Propagation Model

In NS-2, the Propagation models are responsible for calculating the signal-to-noise ratio at the receiver after attenuation and ambient noise are taken into account, as well as the interference range of a signal.

The AquaTools implementation only requires the user to choose the appropriate underwater propagation model in the TCL simulation script using the names for the respective propagation model based on the namesake of the path loss model that forms its basis:

Propagation/UnderwaterThorp

Propagation/UnderwaterFisherSimmons

Propagation/UnderwaterAinslieMcColm

To calculate the signal-to-noise ratio (SNR) at the receiver and the interference range, both the attenuation of the acoustic signal in water and the ambient noise need to be accounted for. The total attenuation is calculated based on the spreading loss, ambient noise and the signal attenuation. The signal attenuation is obtained from either of the Equations 6.2, 6.3 or 6.4 depending upon the path loss model that is chosen as the basis of the underwater propagation model.

The ambient noise in the underwater environment is contributed majorly by four factors; namely, *turbulence, shipping, wind* and *thermal*. The effect of each of these components of ambient noise in the underwater environment may be obtained from Equations 6.5, 6.6, 6.7 and 6.8 which were discussed in Section 6.1.4. A total effect of the noise model may be arrived at by using Equation 6.9 and then converting the obtained value to dB re μPa by using the relationships expressed in Equations B.1 and B.3.

By default the values for the shipping variable, s, and the wind variable, w, are set to 0. These variables are bound to TCL variables called *ship_* and *wind_* respectively and can be set in the usual way with lines such as:

*Propagation/UnderwaterThorp set ship_* **value**

*Propagation/UnderwaterThorp set wind_* **value**

… CHAPTER 7. SOFTWARE IMPLEMENTATION

where *ship_* can take values from 0 to 1 and *wind_* , which represents wind speed, can take positive values in m/s.

Combining the effects of path loss due to absorption and taking into account the spreading loss as well, the total signal attenuation at the receiver is calculated using Equation 6.1. The value obtained here is used in the calculation of the SNR at the receiver in combination with the ambient noise calculation. This calculation is done by a function that overloads the *Pr* function of NS-2 and uses a form of Equation 6.10 to arrive at a result of the received power.

The NS-2 simulator has a node class that keeps information specific to each node in the simulation, including location coordinates (x, y, z) and transmit power settings. The node class also has a number of member functions used to access information about the nodes. The *Pr* function takes pointers to the two communicating nodes and is used by the Channel model in the calculation of packet loss probability. To find the attenuation for a given transmission between two nodes, the center frequency for the transmission must be found. In the implemented model, this corresponds to the frequency that exhibits the best propagation conditions, for a specific distance between the communicating nodes.

As such, in order to obtain this center frequency, the SNR may be expressed as in Equation 6.11 or as a simple function of frequency, which is given by:

$$SNR(l, f) = \frac{P_{tx}}{A(l, f)N(f)} \quad (7.1)$$

This representation makes it clear that the SNR is inversely proportional to the AN factor. As already discussed in Section 6.2.3, the optimal frequencies are those for which the 1/AN value is highest. Keeping this in mind, in order to obtain the center frequency, and thereby the received signal strength, the distance between nodes is calculated. The AN factor for every possible transmission frequency is then calculated and the frequency with the lowest AN factor (largest value of the *AN* variable) is tracked. Finally, the AN factor that corresponds to that frequency is combined with the transmitted power to calculate the SNR at the receiver and is taken to be equal across the the frequency spectrum.

The NS-2 propagation model is also expected to define the radius in which a transmission needs to be considered for interference with other nodes' transmissions. The function *getDist* takes a threshold received power level, the transmit power level and the frequency at which the signal was sent, and returns the largest distance that a node should be from the transmitter and still be considered interfered with by its transmission.

Essentially, this function finds the target attenuation that is needed to result in a received signal strength so low that it does not need to be considered for interference calculations. It then iteratively calculates the attenuation at distances starting at one meter until it finds the target factor. This function is only accurate to the closest meter.

The results obtained by the propagation model are by the channel model to make collision and transmission error decisions. As such, it does not need

CHAPTER 7. SOFTWARE IMPLEMENTATION

to calculate propagation delay or bandwidth. However, these functions are implemented in the channel model, which is described in detail further.

7.1.2 Underwater Channel Model

The channel model in NS-2 maintains the node lists used to calculate neighbor sets, collisions and etc. It is additionally responsible for calculating propagation delays. Essentially, the physical layer calls a *sendUp* function with a packet and a pointer to itself, and the channel model calculates neighbors that may be affected by the transmission as well as propagation delays and returns this information. Full details on the exact functionality of the NS-2 simulator can be found in the NS-2 manual [51].

Aside from calling the appropriate propagation model functions, such as *getDist*, the NS-2 channel model has to implement the propagation delay model as well, which is somewhat complex due to the dependency of the speed of sound on the depth of the water. In addition to the depth in the water, the propagation speed also depends on the temperature and salinity of the water, which in turn depend on the depth through a non-linear relationship. A sample of this non-linear relationship can be seen in Figures A.1 and A.2 of Appendix A.

```
double getTemperature(const double depth) {
    if (depth > 0 && depth < 250)
        return 22;
    else if (depth >= 250 && depth < 750)
        return (22+((250-depth)/35.7));
    else if (depth >= 750 && depth < 1000)
        return (8+((750-depth)/62.5));
    else if (depth >= 1000 && depth < 4500)
        return (4+((1000-depth)/1750));
    else if (depth >= 4500 && depth < 8000)
        return (2+((4500-depth)/875));
    else
        return (-2.00);
}
```

```
double getSalinity(const double depth) {
    if(depth > 0 && depth < 250)
        return (35.5+((0-depth)/833.34));
    else if (depth >= 250 && depth < 500)
        return (35.2+((250-depth)/208.34));
    else if (depth >= 500 && depth < 1000)
        return (34.4-((500-depth)/5000));
    else if (depth >= 1000 && depth < 1500)
        return (34.5-((1000-depth)/2000));
    else if (depth >= 1500 && depth < 4500)
        return (34.75-((1500-depth)/12000));
    else
        return (35.00);
}
```

Figure 7.2: Implementation of the getTemperature and getSalinity functions which provide respective values as a function of depth according to the globally observed average thermocline and halocline.

In order to provide a realistic simulation, the global average observed thermocline and halocline are modelled within the AquaTools implementation as the functions *getTemperature* and *getSalinity*, which provide the temperature and salinity, respectively, as a value which is proportional to the current depth.

With these values obtained the speed of sound can be modelled easily using the relationship defined in Equation 2.1. There are only five known ocean zones where the speed of sound can be expressed as a linear relationship [7], and only for these zones the simulator would not provide results which should be closely matched to reality.

In order to calculate the propagation delay, the *getPDelay* function takes segments of distance traveled depending on the nodes' depth and calculates distance traveled divided by the speed. When all of the segments of the path have been added together, the total propagation delay is returned. A function *SetDistVar* takes the current values of the highest and lowest depth (z-variables) and returns the distance traveled in the next segment of linear temperature change, the average temperature in that zone and the updated values for the z-variables.

To use the underwater channel model, it is only necessary to choose it in the TCL simulation script using the name, *Channel/UnderwaterChannel*.

There is only one bound variable in the channel model that may be set by the user in order to override the *getSalinity* function. The salinity value for the water used in the propagation delay calculation can be set to some other value than the one returned by *getSalinity* as shown below:

$$Channel/UnderwaterChannel\ set\ salinity\_\ \textbf{value}$$

The physical layer model uses information from both the channel model and the propagation model to calculate transmission times, total delays, and the success or failure of packet reception. The physical layer model is described in detail further.

7.1.3 Underwater Physical Layer Model

The physical layer model of NS-2 calculates the final statistics used in the simulation with respect to packet reception, including packet error, transmission time, and propagation delay. For most of these calculations, calls are made to functions in the channel and propagation models. Additionally, information about energy costs associated with the physical interface are stored and used to calculate residual battery charge and transmission energy costs.

All the specific parameters of interface energy consumption are implemented as bound variables to be set by the user, since they depend on the specific hardware being modeled. Additionally, the received signal strength threshold and the maximum transmit power levels are interface specific and are set through bound variables. The default sets of parameters for the maximum transmit power, receive threshold, and the interface energy consumption parameters are set to model the WHOI micromodem [52], since this appears to be a modem that is used most often in academic research due to it's open design and platform. All these parameters can be set up using the normal TCL statements which are used to set up the interface parameters of wireless radio devices in the 802.11 physical layer model as well.

To use the underwater physical model, it is only necessary to choose it in the TCL simulation script using the name:

$$Phy/UnderwaterPhy$$

To set the maximum transmit power and the receive threshold, set the variables *Pt_* and *Pr_* respectively (units in dB re μPa) as shown below:

$$Phy/UnderwaterPhy \text{ set } Pt\_ \textbf{ value}$$

$$Phy/UnderwaterPhy \text{ set } Pr\_ \textbf{ value}$$

The primary function of interest used in the physical layer is the calculation of the available bandwidth given the distance between the transmitter and receiver, their depths and the ambient environmental conditions. Even though the bandwidth calculation function *getBandwidth()* resides in the propagation model it is described here since this is the only place where it is used. First, using the distance between the transmitter and receiver, the frequency experiencing the minimum AN factor is found (optimum frequency). This frequency is used as the center frequency for communication. Then, the 3 dB definition of bandwidth is used to find the edges of the usable frequency band and then bandwidth is calculated as per Equation 6.12 and returned.

7.1.4 Underwater Modulation Model

The Modulation model in NS-2 is responsible for bitrate and bit error calculations based on signal strength and modulation scheme utilized. The error probability is a function of the SNR. The bitrate and number of bit errors is returned by the modulation model.

Even though many modulation schemes have been briefly discussed in Chapter 3, none have been specifically implemented within the AquaTools simulation toolkit since this was presently beyond the scope of this work as it is evident from current literature that not enough investigation work has been done to devise mathematical models that quantify the performance of a particular modulation scheme. As such, the only results available, which are few, are based completely upon experimental observations in off-shore testing and are not numerable enough to formulate mathematical models that could be implemented in the simulator to obtain dependable results. The development of mathematical models that could dependably predict the performance of modulation schemes remains a topic for future research.

As a result, the AquaTools simulation toolkit currently utilizes the wireless modulation scheme as it is provided with an NS-2 distribution in order to perform the bit error calculations. The bitrate utilized is limited to the capacity predicted by an implementation of the Shannon capacity theorem, a mathematical relationship for obtaining which was provided in Equation 6.15.

The modular nature of NS-2 ensures that, when a competent mathematical model for effects of modulation schemes on bit-rates and bit-error rates is developed, it could be easily implemented and replace the current one.

7.2 The USARSim Wireless Simulation Server

USARSim is a high-fidelity simulation tool for simluation robots and environments based on the Unreal Tournament game engine. The software is specifically designed as a research tool and is the basis for important current-day robotic simulations, the most famous being represented by the RoboCup rescue virtual robot competition. The advantages of USARSim consist of the ability to offload the most diffcult aspects of simulation to a high volume commercial platform, which provides superior visual rendering and physical modeling. Therefore, the entire effort can be devoted to the robotics-specific tasks of modeling platforms, control systems, sensors, interface tools and environments. Further advantages that confirm USARSim as a leading robot simulation environment include the presence of development tools integrated with the game engine and advanced editing features for almost every aspect of the simulation, with a special focus on robots and environments. All this functionality allows for a wide range of robot tasks and simulations that can be modeled with greater fidelity in less time. USARSim can also be used to simulate scenarios involving cooperative operation of multiple robots.

All these advantages and its modular nature in developing new additions for sensors, modules and ability to model complex underwater environments makes it a suitable tool to model the multi-AUV underwater acoustic communications as well. This section discusses the underwater environmental and submersible vehicle modelling capabilities of USARSim along with information on the Wireless Simulation Server and the extensions made to these tools in order to enable mobile multi-AUV communication simulations.

Although a part of this work, involving the development of the basic simulation tool was done for a Robotics seminar, the rest of the work of extending the environment to support models besides Thorp and Fisher & Simmons, calculation of bandwidth (not capacity) and the optimal frequency were performed during the course of this investigation in order to leverage the real-time capabilities of this environment.

7.2.1 Underwater Vehicle and Environment Model

In order to correctly evaluate the communication model and test the effects of algorithms, methods and control schemes, it is important to have environment and robot models that mimic reality. USARSim has a model world that simulates an underwater environment available by default, but others can also be easily created using the Unreal Tournament model editor. The default model, shown in Figure 7.3 is used for all the testing associated with the development work.

Along with the underwater world model it is also necessary to have an accurate model of the vehicle to be simulated. This ensure physically accurate simulations of the responses and behavior of a vehicle which in turn assists in accurately testing the performance of an underwater acoustic network. Though any vehicle models

CHAPTER 7. SOFTWARE IMPLEMENTATION

can be created and imported into the USARSim environment, a Submarine model, which is deisgned keeping in mind all physical properties, is provided by default. This model can have sonar sensors, imaging sensors, echo sounders, side scan and an optical camera simulated on it.

Figure 7.3: Screenshot of the USARSim default model and submarine

All the data, including robots, sensors and other devices, associated with USARSim within Unreal Tournament are accessed and manipulated by using TCP/IP socket connections. In order to simplify the interfacing of test scripts and/or software, an interface library developed by the Jacobs University Robotics Group was utilized to spawn connections with the USARSim Simulation Environment.

The default implementation of the interface library for USARSim did not have an implementation of a driving mechanism for the submarine and as such a drive mechanism for the propeller, rudder and stern planes was implemented, thereby providing full mobility control of the submarine and giving access to testing mobile-AUV communications by taking into account real-time decisions and responses made by submersible vehicles.

7.2.2 Wireless Simulation Server

An USARSim plugin called the Wireless Simulation Server (WSS) enables simulation of 802.11 wireless network links within the USARSim environment. WSS works using plugins to implement propagation models allowing for extensibility in the future. The degradation of the arriving signal at the receiving

CHAPTER 7. SOFTWARE IMPLEMENTATION

vehicular node is calculated based upon parameters that are setup for the propagation model plugin and which governs whether connection between robots is possible or not. WSS maintains a table of all USARSim nodes and also information on whether a connection between them, dependent on the propagation model utilized, is possible or not.

Figure 7.4: Screenshot of the USARSim WSS capable of simulating underwater networks

A connection model similar to that of USARSim is utilized by WSS to communicate with all the robot nodes spawned within USARSim. A TCP/IP socket connection is opened by each robot node within the USARSim simulator to WSS in order to query for available nodes, evaluate channel characteristics, check the possibility of connection to another robot and also send data to the remote robot node.

Using this model as a basis the propagation and channel models discussed in Chapter 6 and previous sections were implemented as propagation model plugins for WSS. The model configuration dialog shown in Figure 7.5 shows how the models were designed with flexibility in mind so that the user could configure the ambient noise parameters to suit the real environment being modeled. The model configuration dialog also allows the user to easily configure the signal transmission strength, signal cutoff strength, bandwidth and center frequency in order to easily model any modem available without making changes to WSS or USARSim.

CHAPTER 7. SOFTWARE IMPLEMENTATION

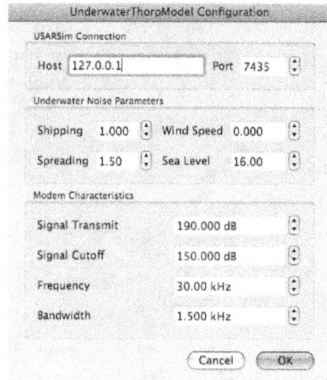

Figure 7.5: Screenshot of the propagation model configuration window

Since USARSim does not have a way to provide the depth of the robot to WSS, a sea level function which defines the sea level in the world map being used was implemented so that the robot's depth could be calculated using its cartesian coordinates, which are available to WSS by directly querying the USARSim server. This depth is utilized in order to determine the temperature and salinity at the current depth from an implementation of the global thermocline and halocline averages in order to compute the propagation delay and attenuation coefficient.

In it's native form, WSS only supports the functions of robots being able to retrieve signal strength from WSS for the target robot and determine whether connection is possible based upon this information and the modem properties. The following commands can be issued over the TCP/IP socket connection in order to retrieve the signal strength information:

GETSS   returns the signal strength at the target robot from the current position of the querying robot.

The unit of the returned data is dB re μPa. However, this limitation is inadequate for the underwater networking scenario where the ability to retrieve propagation delay and channel capacity is also important. As such the following commands were also implemented:

GETPD   returns the propagation delay between the querying robot and the target robot specified in the query string.

GETBW   returns the channel capacity in kbps between the querying robot and the target robot specified in the query string.

These functions utilize the same strategy of calculating the propagation delay and bandwidth as was implemented within the different propagation and channel models for the NS-2 simulator.

A major advantage of the USARSim simulation environment over NS-2, besides the ability to account for realtime robot node reactions to the environment, is that

CHAPTER 7. SOFTWARE IMPLEMENTATION

of being able to successfully emulate an underwater environment. In other words, the environmental modelling ability gives the capability of also modelling and obtaining a surface bottom profile of the ocean floor. This is extremely helpful since the surface bottom of the ocean is a great contributor to signal interference as a result of reflections that occur from the seabed in shallow water acoustic communications. Furthermore, the surface bottom profile can have a significant effect upon multi-path propagation interference as well. As such, it is important to be able to test the likelihood of this factor interfering with the transmission signal. In order to implement the ability to test for surface bottom multi-path signal interference WSS was extended to support the following function as well:

GETML   returns the interference likelihood as 0 or 1 for a distance to the surface bottom provided in the query string.

In order to use the *GETML* function the submarine has to have a scanning sonar mounted. The scanning sonar obtains a bottom profile of distances to the surface and these are supplied to WSS using the *GETML* function. An overall multi-path likelihood using the surface bottom profile supplied by the scanning sonar is calculated by testing each individual acoustic channel pathway caused by reflections from the surface bottom for the arriving signal strength at the receiving robot node. In case the receiving signal strength of any of the surface bottom reflected paths is equal or greater than the cutoff strength of the modem, an interference likelihood of 1 is indicated. This may also be modified to represent the likelihood as a percentage value by calculating the number of total reflections successfully causing interference against those that do not.

7.3 Discussion

As part of this thesis investigation work, the propagation and channel models that were developed in Chapter 6 have been implemented within the framework of a toolkit for the NS-2 networking simulator. This toolkit, named AquaTools, provides access to Propagation, Channel and Physical layer models which are suitable for the underwater acoustic channel, but does not provide a modulation model since currently there does not exist enough experimental information or basic theoretical models from which a mathematical model suitable for this task could be extrapolated.

The implementation of an underwater acoustics channel model in NS-2 provides researchers tools to be able to test multiple different protocols, strategies and methods for tweaking or developing underwater acoustic communication systems. Furthermore, the familiar working environment of NS-2 is retained by AquaTools in the form of setting up experiments easily and quickly via the TCL script interface.

The developed propagation and channel models were also implemented within the framework of the USARSim mobile robotics simulator. In addition to simulating the signal strength, propagation delay, connectivity achievability and channel capacity, the WSS plugin implementation also provides the ability to simulate and provide information on the surface bottom reflective interference as a result of the ability to simulate an underwater environment inherent within USARSim.

The WSS approach provides a significant upper hand over the NS-2 implementation in the form of being able to provide information on surface bottom reflection interference likelihood. Furthermore, the ability to simulate events in a realtime fashion rather than depending upon pre-generated mobility scripts, as is the case with NS-2, can also be viewed as an advantage, especially for scenarios where the mobility pattern of the nodes being investigated cannot be precomputed and may be dependent upon complex environmental interactions. However, this advantage posed by WSS is mitigated since there is no ability to simulate the effects of protocols, modulation schemes and access control by using the WSS approach. On the other hand NS-2 provides the advantage of access to these abilities and also being able to simulate complex network behavior such as collisions and multiple-access interference.

Chapter 8

Simulator Validation

Both simulators developed as part of this work, the AquaTools NS-2 Toolkit and the USARSim WSS plugin, provide generic tools to the underwater acoustics communication community in order to test and develop underwater acoustic communication systems. Having focus upon two of the largest different user groups for such systems, one networking and the other robotics, these simulation environments provide not only tools that would be very valuable but also those which are within frameworks familiar and often used within these communities. However, before any simulation tool can be utilized to take dependable design decisions, it is required to validate the results obtained from the simulator in order to ensure that they confom to those that are available within published literature or they conform to those expected from numerical models utilized and published in literature.

The numerical evaluation of the propagation and channel models carried out in Chapter 6 already establishes the initial soundness of the models. Furthermore, since the models which form the basis of the overall propagation and channel models are widely used and published in academics, they can be viewed as dependable. In fact, similar design methodology followed in this work has already been utilized in other published work [8, 7, 53], thereby pointing towards the soundness of the mathematical models used. The results achieved by the numerical analysis are in line with those expected from published literature [8, 53].

Even though this points towards a sound mathematical model for the propagation and channel models, mathematical verification only supports the case for their utilization in the simulator. The results provided by the simulator also need to be validated in order to ensure that these are within expected margins of the numerical models and closely mirror those already in literature.

Only upon successful validation can the simulators be used to set up experiments which can be utilized to derive results that can aid in development of technologies supportive of underwater acoustic communications. To validate the implemented underwater models in both the simulators, a number of simulations were run and the resulting values of specific parameters compared with those calculated using analytic models and published literature, where available.

CHAPTER 8. SIMULATOR VALIDATION

Specifically, it was considered important to validate the major characteristics of the simulator; namely, noise, AN factor, optimal frequency, propagation delay, bandwidth and capacity to ensure there were no errors in the implementations. As will become evident from the presented results, the results obtained from both the simulators matched the analytical model and also the published results which were available for comparison.

Both simulators are designed with the same basic propagation and channel models and consequently the results from one simulator can easily be obtained in the other simulator as well. As such, for all the verification cases, the presented results are from one of the simulators, but tests were carried out in both simulators to ensure accuracy of results.

8.1 Noise

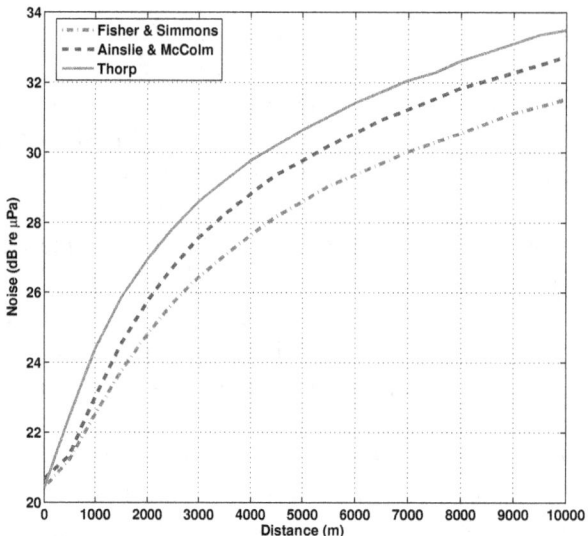

Figure 8.1: The changing ambient noise as per changing distance which effects the optimal frequency used for noise calculation.

The noise calculations are critical for calculating all the important parameters of the propagation and channel model, such as bandwidth, capacity and SNR. As such it was considered vital to evaluate the accuracy of the simulators in calculating the noise.

The noise predicted as per all the three models is plotted in Figure 8.1. The optimal frequency for transmission is used to arrive at an estimate of the ambient noise since the optimal frequency provides the best case performance. The results

CHAPTER 8. SIMULATOR VALIDATION

shown here are from the AquaTools NS-2 Toolkit, however, the results produced by the USARSim WSS environment were exactly the same. The minor differences in the shape of the curves can be attributed towards the fact that each model accounts for different environmental parameters and as it is already known from Chapter 6 that the Ainslie & McColm model outperforms the other two models which provide the upper and lower bound in this case.

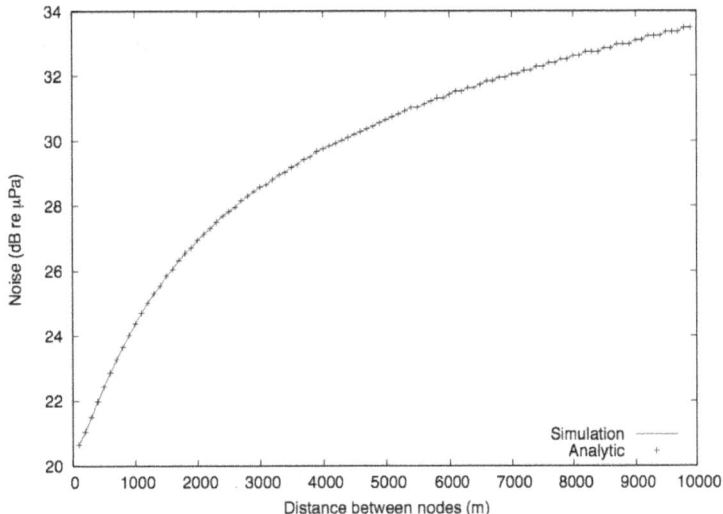

Figure 8.2: The ambient noise as obtained by the simulative and analytical study conducted by Harris *et al.* while using the Thorp model [7].

While the noise is not dependent on the transmission distance, verifying this result is necessary since ambient noise is dependent upon the frequency used for transmission and thereby, here it indicates whether the optimal frequency predicted by the used models is accurate or not. When compared to the already published results of ambient noise within the work performed by Harris et al. [7], the results of which can be seen in Figure 8.2, we can easily notice that the curves are very similar. In fact, upon close scrutiny, it is evident that the results predicted by the simulator while using the Thorp model are exactly the same as those predicted in published literature [7].

While there is no direct comparison available to verify the results from the other two models, the results available are within expectations. This combined with the accuracy of the simulators while using the Thorp model argues in favor of the accuracy of the models and the simulation environments.

8.2 Propagation Delay

Testing the accuracy of the propagation delay calculation requires a number of experiments since the result depends on the depth of the communication in the water. The test cases used in the simulator utilized two nodes, with one transmitting data to the other. The depth of both the nodes were also varied while simultaneously changing the distances. The results that were obtained appeared to be within expected parameters since the shape of the propagation delay should closely mirror that of the halocline and thermocline models being utilized [54].

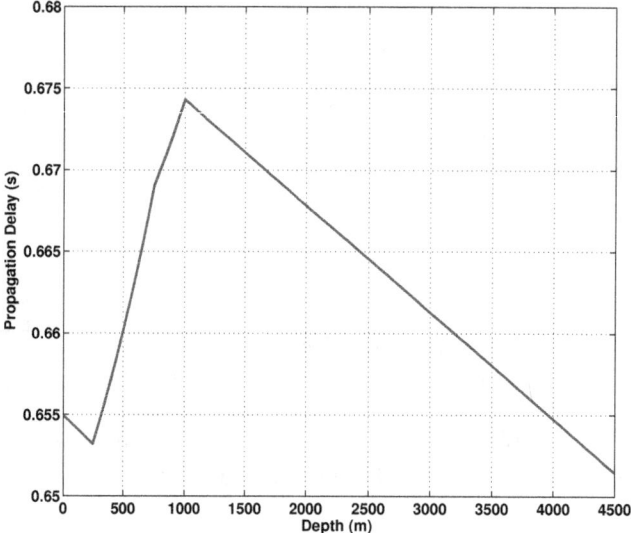

Figure 8.3: The change in propagation delay with depth of the two nodes. The propagation delay curve follows a shape similar to that of the sound velocity profile.

As such, in order to predict the accuracy of the simulator with a degree of certainty, an experiment similar to the one run by Harris et al. [7] was executed with two nodes, both situated 1 km apart. The depth of both these nodes was progressively incresed while maintaining the same depth for both the nodes and keeping the 1 km distance between them constant. The resulting values of propagation delay are plotted in Figure 8.3.

CHAPTER 8. SIMULATOR VALIDATION 65

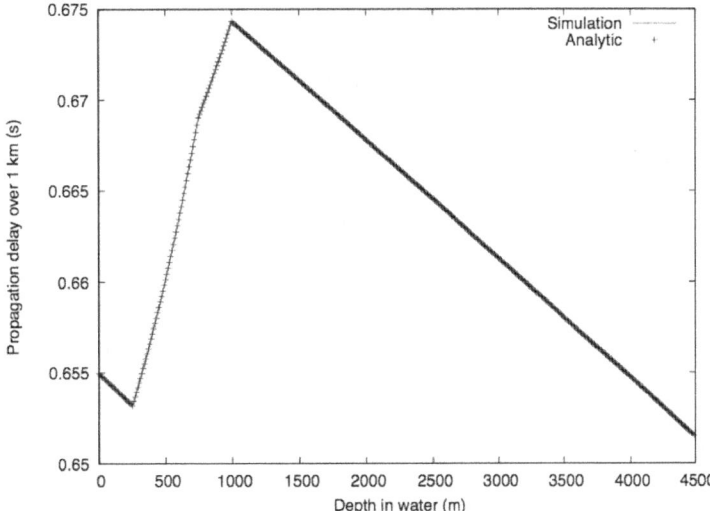

Figure 8.4: The Propagation Delay as obtained by the simulative and analytical study conducted by Harris *et al.* [7].

A direct comparison between the results obtained from USARSim WSS and previously published literature can be had by comparing Figure to Figure 8.4. It is clearly evident that the results obtained from AqualTools and USARSim mimic those in previously published literature, thereby further strengthening the case for the accuracy of these simulation tools.

8.3 Signal-to-noise Ratio

The SNR is an important value that not only assists in choosing modems that might function within a specific network design, but also assists in ensuring that nodes in a network are distributed such that a high network efficiency is maintained. This is so since the SNR determines whether the arriving signal at the receiver has a strength strong enough to be accepted or discarded.

Figure 8.5 depicts the SNR as obtained during the study conducted by Caiti et al. [55] to characterize the underwater acoustic channel. In this study, focus was placed specifically upon the actual operational capabilities of acoustic modems which are currently available. As such, the frequency range of the plot depicted in Figure 8.5 is limited to this operational range.

Furthermore, in their study Caiti et al. considered three cases for testing the operational scenarios; a brief overview of these three scenarios can be found in Figure 8.5. In Figure 8.6 the large black dots represent the transmitter and receiver, whereas the red line represents the shape of the thermocline. As such, by testing for different relative locations of the transmitter and receiver and also accounting for different shapes of the thermocline, which leads to changes in

CHAPTER 8. SIMULATOR VALIDATION

attenuation and other factors, they obtain an overview of the behavior of SNR. The results they obtain indicate that taking the thermocline into account has an effect upon the predicted SNR, however, the effect of the relative location of the receiver and transmitter is not important in most important scenarios while determining the SNR. As such, while choosing to perform an experiment in the AquaTools NS-2 Toolkit, the effect of the thermocline was considered, however, the depth of both the nodes was maintained at a constant of 1 km.

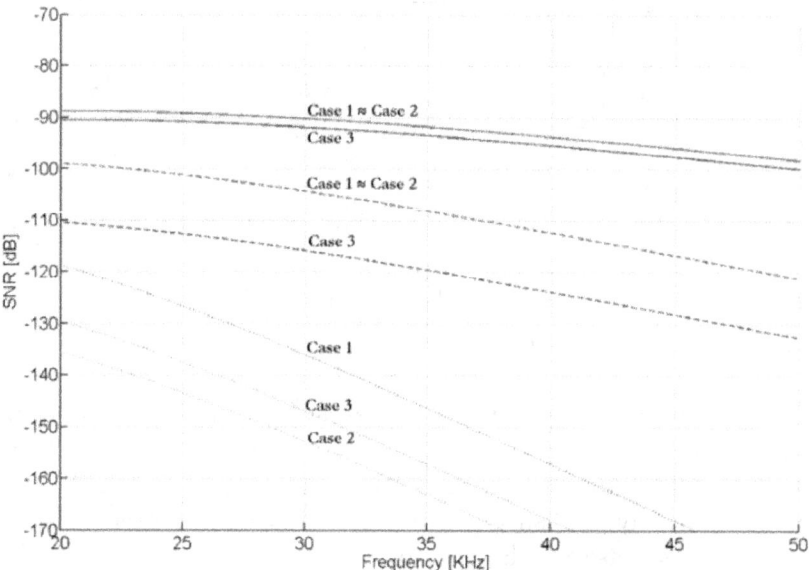

Figure 8.5: The SNR as predicted during the study conducted by Caiti *et al.* while characterizing the underwater communication channel. (Solid lines - 1km, Dashed lines - 2km and Dotted lines - 5km; Three different cases are different operational cases with different transmission powers. Thorp model was used for the study) [55].

The authors of [8] point out successfully in their study that SNR is also closely related to the AN factor which assists in deriving the optimal frequency, bandwidth and capacity. Their work is further extended by Harris et al. [7] who determine that the AN factor is a much better method of generally predicting the performance of the SNR since the values of SNR are specifically determined by the transmission power of the acoustic signal, whereas the AN factor only depends upon the distance and frequency of transmission. However, it is also pointed out in their work that the shape of the AN factor curve would be similar to that of the SNR; the variation can always be accounted for due to the chosen transmission frequency.

CHAPTER 8. SIMULATOR VALIDATION

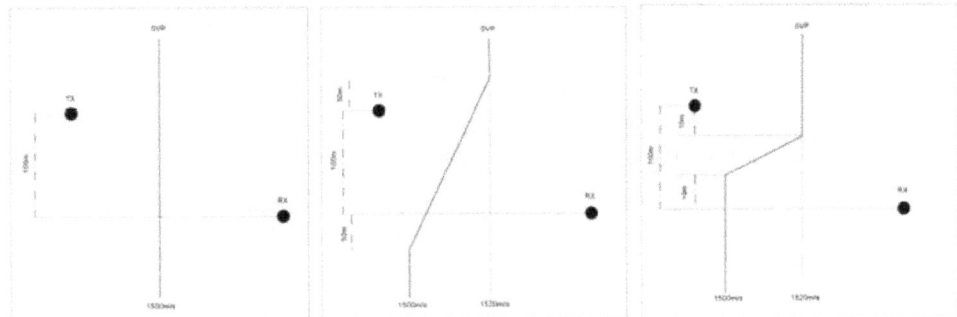

Figure 8.6: The operational scenarios used in the investigation performed by Caiti et al. while characterizing the underwater acoustic channel in operational scenarios (the black dots are the transmitter and receiver pair, whereas the solid red line represents the thermocline) [55].

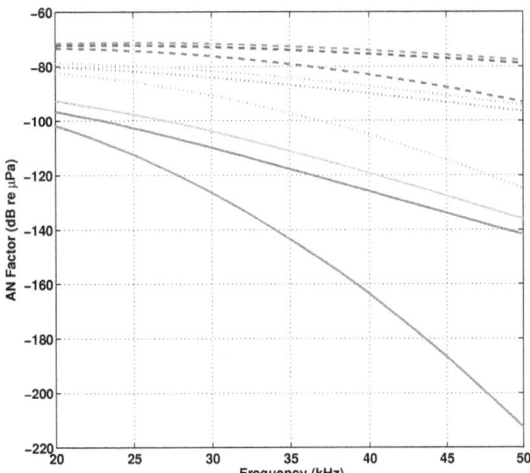

Figure 8.7: The AN factor's relationship with the transmission frequency being utilized. The close relationship with SNR makes AN factor useful to judge performance. Only common operational frequencies are used here. (Dashed lines - 1km transmission distance, Dotted lines - 2km transmission distance & Solid lines - 5km transmission distance; Red - Thorp, Green - Fisher & Simmons, Blue/Gold - Ainslie & McColm)

As such, the AN factor was chosen as the benchmark parameter to validate the performance of the simulators. The results of running experiments with transmission distances of 1 km, 2km and 5 km while using all three models, due to it being proven as more accurate previously in the course of this investigation, are

CHAPTER 8. SIMULATOR VALIDATION

plotted in Figure 8.7. Close examination reveals that the curves here are similar to the SNR curves obtained by Caiti et al. and also resemble the AN curves reported by Harris et al. and Stojanovic et al. in their respective works. The results depicted here are from the USARSim WSS simulation environment, however, the same results were also obtained from the AquaTools NS-2 Toolkit.

Strong similarity between the previously reported and obtained results builds confidence in the accuracy of both the simulation environments and the models being utilized.

8.4 Signal Strength

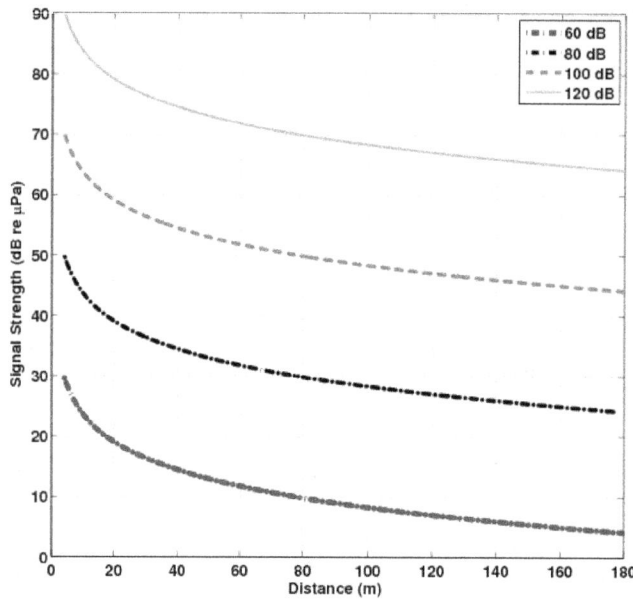

Figure 8.8: The arriving signal strength as predicted by the Ainslie & McColm model while the distance between the transmitting and receiving nodes was varied between 4 to 180m and the transmit power is also changed.

Even though the arriving signal strength does not have a direct influence on the performance of the acoustic channel, it is very useful in determining the quality of the arriving signal and can therefore be used to develop protocols that provide quality assurance, or even simply to choose the appropriate acoustic modem. It can even be utilized to develop adaptive modems that could change the transmission frequency to achieve a target signal strength at the receiver end.

CHAPTER 8. SIMULATOR VALIDATION

The evaluation of the arriving signal strength is not a straightforward comparison like other values since it is dependent upon the transmission signal strength and most of this work in published literature is based upon the transmission strength necessary to achieve a desired SNR level at the receiver. As such, in order to test the accuracy of the simulator in this area, it is important to draw a few inferences from the data that is available thus far.

It is known to us from previous work [53] that the available capacity drops with distance and to achieve a higher capacity higher transmission power is necessary. Conversely, available capacity is proportional to the transmission power utilized. Extrapolating from this information, the equations to calculate arriving signal strength and previously reported results [8, 53], we can easily deduce that the signal strength should reduce with distance in a somewhat logarithmic fashion. Furthermore, as we saw in the previous section, the SNR increases with distance and as such the arriving signal strength must also necessarily reduce with distance.

Keeping this in mind an experiment while keeping a depth of 100 m constant, using the standard thermocline and halocline with the Ainslie & McColm model, and varying the distance between the two nodes between 4 m and 180 m and also changing the transmission power, was executed in the USARSim WSS simulation environment. The results of this experiment can be seen in Figure 8.8. The shape of this figure absolutely follows the expected shape and also shows that capacity increases with higher transmission strength, thereby indirectly validating the models used in the simulators.

8.5 Bandwidth and Capacity

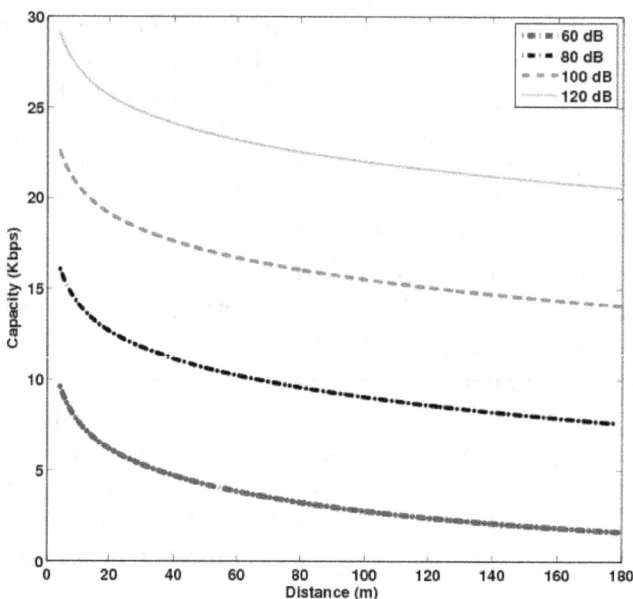

Figure 8.9: The channel capacity as predicted by the Ainslie & McColm model while the distance between the transmitting and receiving nodes was varied between 4 to 180m and the transmit power is also changed.

Finally, the model needs to accurately predict the available bandwidth and the maximum capacity of the channel given the distance between nodes, their depth and also the ambient environmental parameters.

The relationship between bandwidth and capacity is a well established one. Higher bandwidth leads to higher capacity; in fact, the relationship between both these operational parameters is so strong that the curve of a plot of each of these would look identical, as is already shown in the work by Stojanovic et al. [8]. As such, it was considered necessary to only test the performance of the simulators in one of these categories since it would also accurately show the performance of the simulators with concern to the other parameter.

CHAPTER 8. SIMULATOR VALIDATION

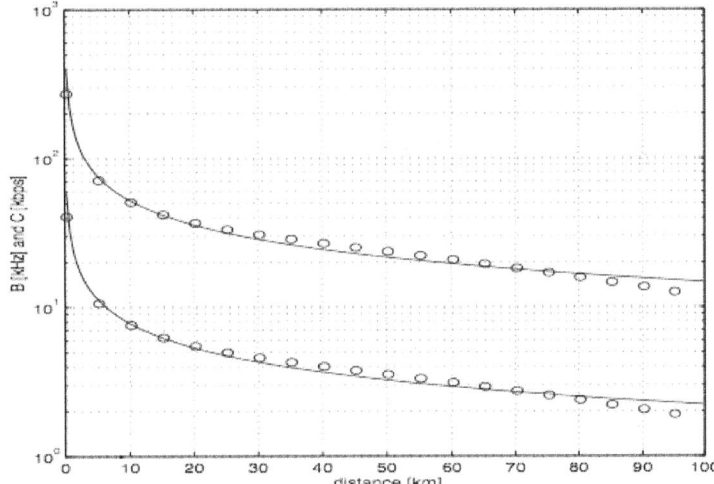

Figure 8.10: The bandwidth and capacity as predicted by the Thorp model while the distance between the transmitting and receiving nodes was varied during the study conducted by Stojanovic *et al.* [8] (Upper line is capacity).

The experiment was set up in the USARSim WSS simulation environment since both the simulators use the same approach for calculating the bandwidth and capacity. Furthermore, since all the building blocks used to calculate these parameters have tested to work properly in both the simulators, there is no problem expected in calculation of these either. For the purpose of the simulation two nodes were created at a depth of 100 m in the environment and the WHOI micromodem was modelled as the acoustic modem of choice. Using different transmission powers within the capabilities of the modem, the results were derived by varying the distance between the two nodes between 4 m and 180 m. Standard parameters defined earlier in this document were used to define the ambient environment and the thermocline and halocline were also taken into consideration.

The results of the experiment can be seen in Figure 8.9. It is clear from this figure that the capacity reduces with distance between the nodes. If the obtained result is compared to the one arrived at by Stojanovic et al. [8] in their work, depicted in Figure 8.10, it becomes clear that the shape of the curves is very similar irrespective of the transmission frequency utilized. Even though the transmission distance used between the nodes is not within the same range as depicted in the results of the Stojanovic et al. work, the similarity in the shape of the curves argues in the favor of the overal robustness of results provided by the simulators. Furthermore, the fact that the shape of the curves is similar for all the different transmission powers tested, and even resembles the transmission power curves in Figure 8.8, clearly shows that the simulators are providing results which are dependable.

8.6 Discussion

From all the different parameters that have been tested and compared to previously published numerical, analytical and simulative results, it is quite clear that the models being utilized to simulate the underwater acoustic channel are performing quite well. Furthermore, the results obtained from both the implemented simulation environments are also extremely accurate.

Coupled with the usability of the simulators, the validity of these easily verifiable results makes a strong case for the use of these simulators as tools in further study of the underwater acoustic communication channel.

Chapter 9

Conclusions & Future Directions

During the course of this investigation multiple numerical models were formulated and then analysed by calculating results that provided an insight into the behavior of the underwater acoustic communication channel. In order to obtain a deeper understanding of the overall functioning of the underwater acoustic communication channel, two different simulators with different capabilities were implemented as tools for the broader underwater acoustic networking community to use as aids in conducting research and developing protocols or tools to assist in reliable underwater acoustic communication.

Owing to the wide range of data that has been explored as part of this investigation, this chapter will provide a short overview of the overall work that was performed and also the resulting conclusions that can be concretely drawn in respect to the underwater acoustic communication channel. Some future directions that can be followed based upon this work will also be highlighted.

9.1 Contributions

The aim of this investigation was to develop one or more mathematical models that would accurately and as completely as possible describe the underwater acoustic channel. Furthermore, after development of the mathematical model it was considered important to develop simulation tools that would assist further in the investigation of the underwater acoustic channel so as to be able to arrive at an overall understanding of this highly complex and dynamic communication medium.

As part of this thesis investigation work, the following contributions have clearly been made:

1. Existing mathematical models that define the properties of the underwater acoustic communication medium were combined to develop an overall propagation and channel model that completely characterizes this medium. Furthermore, the channel model was developed in such a way that it can

be applied to numerical analysis or even software based simulations without any modifications.

2. Three independent channel models were developed in order to extend the understanding of the underwater acoustic channel, as against the availability of only one model thus far. The existing models only provided access to the ability of being able to model the effects of distance and transmission frequency on the channel, however, the new channel models developed in this investigation also add the ability to model the effects of depth, temperature, salinity and acidity of the ocean. Keeping in view the dynamism of the underwater environment, a complete understanding of the effect of all these parameters is also critical and hence highlights the importance of having channel models that consider these as well.

3. Two simulation environments were developed in order to offer a simulation toolset, as complete as possible, to the wider underwater acoustic communication community. Together these simulation tools offer the ability to simulate not only the physical effects of the communication channel but also networking specific performance criterion like packet collisions, media access and others. The simulators also provide the ability to investigate the effects of pre-known mobility patterns or even perform simulations in real-time in order to observe the performance of a network on mobile underwater vehicular nodes.

4. A clearer understanding of the effects of depth, temperature, salinity and acidity on underwater acoustic communications is also now available.

9.2 Conclusions

As a result of the experiments that were carried out as part of this investigation, a number of conclusions regarding the underwater communication channel may now be drawn. An overview of these conclusions are presented below:

1. The many factors, such as temperature, salinity, acidity, shipping caused noise and others, effecting the underwater environment cause the acoustic communication channel to become a highly complex one which is highly dynamic and ever changing. As a result, there is no single definable underwater acoustic communication channel, but rather there are many versions of this which are specific to the local oceanic zone and time of day since slight fluctuations in some of these parameters as a result of something as temporary as weather can lead to completely different performance characteristics. This makes it important to accurately characterize the communication channel and predict the performance of a network in order to maximize performance by levariging the advantages available at any

CHAPTER 9. CONCLUSIONS & FUTURE DIRECTIONS 75

given time. This further indicates that it is impossible to use a single acoustic modem and achieve the best possible performance all across the global oceans. In order to achieve this ability, it would be necessary to investigate and develop adaptive modems that are capable of sensing the ambient environment, characterizing the channel based on this and then modifying their performance based on these factors.

2. The different factors characterizing the channel performance are many, however, the best parameter to measure performance of a channel is the maximum capacity. The underwater acoustic channel's capacity is effected by a host of different parameters. Unlike terrestrial networks, it was already known that the transmission distance and frequency both have an effect on the available capacity. It has been shown in previous work that increasing transmission distance reduces available capacity. However, through the course of this investigation it has also been discovered that greater depths increase channel capacity, while operating in higher temperature, acidity and salinity also increases the capacity of a network.

3. The attenuation coefficient determines the overall signal attenuation that occurs over a certain distance. The relationship between transmission frequency and the attenuation coefficient is well understood, however, this study indicates that the attenuation coefficient not only increases with the used transmission frequency, but decreasing temperature, salinity and acidity as well. On the other hand, increasing depth causes the attenuation coefficient to decrease, therby increasing the likelihood of a signal with high strength to arrive at the destination if deeper depths are utilized for acoustic communications.

4. The optimum frequency is that at which the best channel bandwidth and capacity is available. As a result, it is preferable to operate at this frequency and this makes it important to have an understanding of this as well. The optimum frequency reduces logarithmically with the transmission distance, thereby making it better to transmit data over longer distances since higher frequency transmissions generally amount to higher battery consumption, which is a bane in underwater communications as most nodes are equipped with limited power supply. In fact, the operational features of most modems make it appropriate to use them between a range of about 10 m to 5 km. On the other hand, temperature and depth cause the optimum frequencies to increase. However, this increase is within the operational range of most modems and as such becomes less important to account for.

5. It has been noticed in previously reported results and also during the course of this investigation that the relative positioning of the nodes in the underwater environment has a pronounced effect on some of the network and channel performance characteristics, whereas others remain uneffected.

For example as seen in previous chapters, the SNR is not effected by the relative positioning of the nodes in the network in the depth plane, however, the distance between the nodes does effect this. This is of great importance while choosing a network design in accordance with a modem or a modem for a particular network design. However, depth of the nodes becomes important when considering the capacity of the channel. The same is applicable for the ambient temperature, salinity and acidity, even though the effects of salinity and acidity on the overall capacity is neglegible.

6. It is shown in the previous point that the relative position of the nodes may not have a pronounced effect on all the parameters characterizing the channel or network performance. However, unlike the positioning of the nodes, the halocline and thermocline of the oceanic region being considered has a pronounced effect on all of the channel and network performance characteristics, thereby making it important to ensure that any simulation or numerical analysis undertaken takes these into account as well.

9.3 Future Directions

As a result of the investigation that has been performed as part of this thesis work, a few research directions have been identified for the future. Some of the important research directions are listed below:

1. Since there is no mathematical model currently available that characterizes the effects of different modulation schemes on the performance of underwater acoustic channel based communications, it would be important to find more experimental data from which such a model could be designed. This would be extremely helpful in further enhancing the current simulators by allowing them to not only calculate the theoretical maximum possible bitrates, but also the maximum achievable bitrates as a direct effect of the coding and modulation schemes used. This would make simulations more realistic and ensure a much higher accuracy of results.

2. The current noise models are adequate for a rough approximation, however, for deep sea operations they are not completely appropriate since they do not attenuate as a direct effect of depth. At the moment, they are designed to attenuate with depth only since the optimal frequency chosen at greater depths is lower, leading to lower attenuation due to ambient noise. Extending these models based upon depth based attenuation, perhaps by using one of the absorption coefficients, would be extremely useful.

3. It has been highlighted during this investigation that shorter distances and deeper paths provide higher network capacities. As such, it would be useful to investigate the effects of designing a routing protocol that could utilize

these benefits to maximize the capacity of the network, thereby, possibly delivering the data at the highest achievable data rate.

Part IV

Appendices

Appendix A

Characteristics of Sound Velocity Parameters

A.1 Ocean Temperature Profile

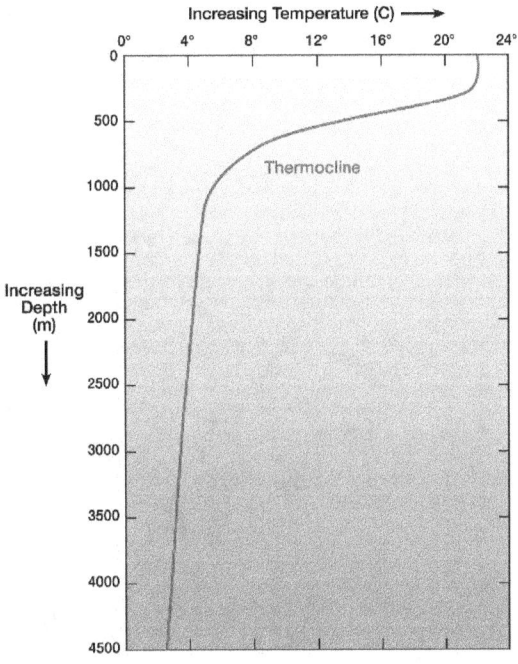

Figure A.1: Ocean water temperature with depth [54]

Most of the light and heat radiated on to the ocean gets absorbed within the first few tens of meters of water but wave and turbulence cause the heat to be transferred to lower layers of the ocean rather quickly. As seen in recorded data, the temperature of the surface waters varies mainly with latitude. The polar seas can be as cold as -2°C while the Persian Gulf can be as warm as 36°C and the average temperature of the ocean surface waters is about 17°C [54].

The boundary between surface waters of the ocean and deeper layers that are not mixed is termed the thermocline and it usually begins at around depths of 100-400 m and extends several hundred of meters downward from there. As shown in Figure A.1 the temperature in the thermocline region drops rapidly and as such makes it important to have accurate measurements available for this region. Below the thermocline region temperatures approach 0°C with a steady downwards curve for the temperature curve, as seen in Figure A.1.

A.2 Ocean Salinity Profile

Ocean salinity needs to be analyzed in both, surface and depth profile. This section provides insight into the ocean salinity profile.

A.2.1 Salinity-Depth Profile

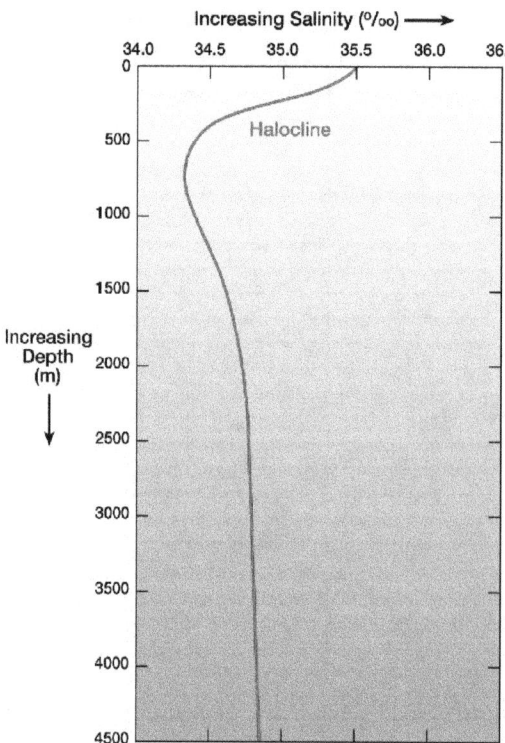

Figure A.2: Salinity-depth profile for South Atlantic Ocean [56]

As shown in Figure A.2, from the surface to the deeper layers salinity of the ocean water varies between 33−37 ppt [56]. Generally, the salinity of surface ocean water is high and then decreases until a depth of about 1000 m, within the halocline layer where the mixing layers of water cause the salinity to change rapidly. Below this range the salinity of ocean water once again starts to increase with depth at a slow rate. However, at no point the measured salinity crosses 37 ppt and as such an average ocean salinity of 35 ppt for depth applications is a good measure to adopt.

A.2.2 Surface Salinity Profile

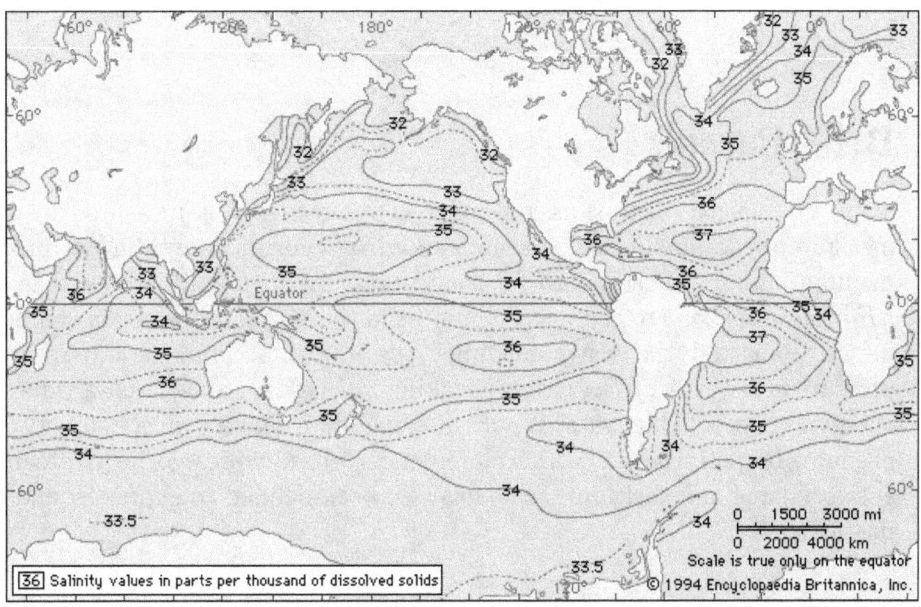

Figure A.3: Average global ocean surface salinity [57]

Figure A.3 shows the salinty of ocean surface water as measured globally. It is clear from the image that the range of salinity values for the ocean surface waters corresponds to a salinity value between 33-36 ppt [57]. As such, the average surface salinity for ocean water is about 34.5 ppt. Using this information along with the the average ocean water salinity from the previous section shows us that the overall average variance for ocean water salinity is within the range of 33-37 ppt, thereby making 35 ppt a seemingly acceptable average irrespective of depth and lattitute or longitude.

Appendix B

Sound Energy Units

B.1 Pascals

Typically, acoustic transmission and reception energy is measured in μPa as a measure of the amount of pressure caused by acoustic waves in the propagation medium. In air, the standard for sound pressure is globally accepted to be as 20 μPa. However, due to the vast differences in the sound propagation characteristics between air and the water medium, the global accepted standard for sound pressure in water is 1 μPa, thereby making it necessary to not that a value of sound energy as expressed in water is not the same in air and a conversion between these needs to be performed. Furthermore, since values expressed in μPa can tend to get very large, a logarithmic scale has also been defined to express sound pressure in water.

B.2 Decibels

Decibels (dB) measure the sound energy on a logarithmic scale, in order to make it easier to express these values. The dB value can be obtained from a μPa using the following equation:

$$dB = 10 \log_{10} \left(\frac{p}{F_{pr}} \right)^2 \tag{B.1}$$

where, p is the sound pressure expressed in μPa and F_{pr} is the standard sound pressure in the propagation medium.

Equation B.1 may further be extended to be applicable in air and water as below:

$$dB_{air} = 10 \log_{10} \left(\frac{p}{0.00002} \right)^2 \tag{B.2}$$

$$dB_{water} = 10 \log_{10} \left(\frac{p}{0.000001} \right)^2 \tag{B.3}$$

Appendix C
NS-2 Sample Scripts

Though NS-2 provides a familiar networking simulation environment, a few sample scripts are provided along with a description in this chapter in order to aid in the development of more advanced networking scenario simulation scripts. Being an NS-2 toolkit, the AquaTools environment supports the use of all external NS-2 tools to generate mobility scenarios and connection patterns as well. Furthermore, all existing routing protocols and other developments can also be utilized.

C.1 Sample 1 - Static Nodes

```
#================================
# Define options
#================================
set val(chan)           Channel/UnderwaterChannel
set val(prop)           Propagation/UnderwaterThorp
set val(netif)          Phy/UnderwaterPhy
set val(mac)            Mac/802_11
set val(ifq)            CMUPriQueue
set val(ll)             LL
set val(ant)            Antenna/OmniAntenna
set val(ifqlen)         50
set val(nn)             3
set val(rp)             DSR

Phy/UnderwaterPhy set CPThresh_ 10.0;
Phy/UnderwaterPhy set CSThresh_ 0.284;
Phy/UnderwaterPhy set RXThresh_ 4.0;
Phy/UnderwaterPhy set Pt_ 97;
Phy/UnderwaterPhy set freq_ 30;
Phy/UnderwaterPhy set L_ 1.0;
```

APPENDIX C. NS-2 SAMPLE SCRIPTS

```
# ================================
# Setup Trace
# ================================
set ns_           [new Simulator]
set tracefd       [open underwatertrace.tr w]
set nf [open nam-simple.nam w]
$ns_ use-newtrace ;
$ns_ trace-all $tracefd
$ns_ namtrace-all-wireless $nf 15 15

# ================================
# Setup Topography and Create God Process
# ================================
set topo          [new Topography]
$topo load_flatgrid 15 15
create-god $val(nn)

# ================================
# Create Nodes
# ================================
$ns_ node-config -adhocRouting $val(rp) \
                 -llType $val(ll) \
                 -macType $val(mac) \
                 -ifqType $val(ifq) \
                 -ifqLen $val(ifqlen) \
                 -antType $val(ant) \
                 -propType $val(prop) \
                 -phyType $val(netif) \
                 -channelType $val(chan) \
                 -topoInstance $topo \
                 -agentTrace ON \
                 -routerTrace ON \
                 -macTrace ON \
                 -movementTrace OFF

for {set i 0} {$i < $val(nn) } {incr i} {
      set node_($i) [$ns_ node]
      $node_($i) random-motion 0;
}
```

```
# ================================================
# Set Node Positions
# ================================================
$node_(0) set X_ 5.0
$node_(0) set Y_ 5.0
$node_(0) set Z_ 0.5

$node_(1) set X_ 6.0
$node_(1) set Y_ 5.0
$node_(1) set Z_ 0.5

$node_(2) set X_ 5.5
$node_(2) set Y_ 5.0
$node_(2) set Z_ 0.5

# ================================================
# Setup Traffic Flows
# ================================================
set udp [new Agent/UDP]
$udp set fid_ 1
set sink [new Agent/LossMonitor]

$ns_ attach-agent $node_(0) $udp
$ns_ attach-agent $node_(1) $sink

$ns_ connect $udp $sink

# Creating CBR Traffic
set cbr [new Application/Traffic/CBR]
$cbr set packetSize_ 1
$cbr set interval_ 10.0
$cbr attach-agent $udp
$ns_ at 0.0 "$cbr start"

# ================================================
# Simulation Startup and Shutdown
# ================================================
for {set i 0} {$i < $val(nn) } {incr i} {
    $ns_ at 250.0 "$node_($i) reset";
}
$ns_ at 250.0 "stop"
$ns_ at 250.50 "puts \"NS EXITING...\" ; $ns_ halt"
proc stop {} {
```

APPENDIX C. NS-2 SAMPLE SCRIPTS

```
        global ns_ tracefd nf
        $ns_ flush-trace
        close $tracefd
        close $nf
        exit 0
}
puts "Starting Simulation..."
$ns_ run
```

The script shown above is a very simple one that simulates a network of three underwater nodes, the positions of which are fixed. The purpose of this script is to generate CBR traffic which is delivered to every node which is within reception range.

In the "Define Options" section of the script we can see that the Underwater channel is chosen, followed by the Thorp model for propagation and the Underwater Physical layer model for the interface. Since there is no separate MAC interface defined for underwater environment within AquaTools, the 802_11 MAC layer is chosen. As a result, all the other configuration options remain the same as they would for the 802_11 MAC layer. As can also be seen, three nodes are defined in the network, along with the DSR routing protocol chosen for routing decisions. The DSR routing protocol can be replaced with any protocol that is developed for NS-2.

Following the basic configuration of the channel, the physical layer is configured to match the capabilities of the modem. The *CPThresh_* refers to the capture phenomenon, i.e., if two packets are received simultaneously it is still possible to receive the stronger packet if its signal strength is *CPThresh_* times the other packet. In this case, the stronger packet in a collision can be decoded if its signal strength is at least 10 dB times greater than that of the other packet; otherwise both the packets are lost.

The *CSThresh_* is the carrier sensing threshold. If the received signal strength is greater than this threshold, the packet transmission can be sensed. However, the packet cannot be decoded unless signal strength is greater than *RXThresh_*. In this script, the *CSThresh_* is 0.284 dB times lesser than *RXThresh_*. The RXThresh_ is the reception threshold. If the received signal strength is greater than this threshold, the packet can be successfully received. In this script, this threshold is set to be 4 dB. As such, if any packet below a 4 dB strength arrives then it is discarded. The transmission frequency is set to 30 KHz and transmission strength of 97 dB is used.

The "Setup Trace" section creates Tcl variables to store the network trace data in to. This section is not discussed in detail here as this is the standard NS-2 trace configuration method. Following this a topography map is loaded and the God process of NS-2 initialized. The nodes are then created along with all the options that were initialized at the beginning of the script.

It is required that all nodes in the simulation have an initial position recorded. As such, the position of all the three nodes is set up in the "Set Node Positions" section of the script. Here, the Z value defines the depth of the nodes in km. No mobility patterns are created in this script since the nodes are setup in a static topology. All the sections following this in the script are once again standard NS-2. It is clear from this sample script that in order to use the AquaTools toolkit no special knowledge is necessary since simple NS-2 scripts can be written in the familiar Tcl environment.

C.2 Sample 2 - Mobile Nodes

```
# ==========================================
# Define options
# ==========================================
set val(chan)           Channel/UnderwaterChannel
set val(prop)           Propagation/UnderwaterThorp
set val(netif)          Phy/UnderwaterPhy
set val(mac)            Mac/802_11
set val(ifq)            CMUPriQueue
set val(ll)             LL
set val(ant)            Antenna/OmniAntenna
set val(ifqlen)         50
set val(nn)             3
set val(rp)             DSR

Phy/UnderwaterPhy set CPThresh_ 10.0;
Phy/UnderwaterPhy set CSThresh_ 0.284;
Phy/UnderwaterPhy set RXThresh_ 4.0;
Phy/UnderwaterPhy set Pt_ 97;
Phy/UnderwaterPhy set freq_ 30;
Phy/UnderwaterPhy set L_ 1.0;

# ==========================================
# Setup Trace
# ==========================================
set ns_         [new Simulator]
set tracefd     [open underwatertrace.tr w]
set nf [open nam-simple.nam w]
$ns_ use-newtrace ;
$ns_ trace-all $tracefd
$ns_ namtrace-all-wireless $nf 15 15


# ==========================================
```

APPENDIX C. NS-2 SAMPLE SCRIPTS

```
# Setup Topography and Create God Process
# ==============================================
set topo            [new Topography]
$topo load_flatgrid 15 15
create-god $val(nn)


# ==============================================
# Create Nodes
# ==============================================
$ns_ node-config -adhocRouting $val(rp) \
                 -llType $val(ll) \
                 -macType $val(mac) \
                 -ifqType $val(ifq) \
                 -ifqLen $val(ifqlen) \
                 -antType $val(ant) \
                 -propType $val(prop) \
                 -phyType $val(netif) \
                 -channelType $val(chan) \
                 -topoInstance $topo \
                 -agentTrace ON \
                 -routerTrace ON \
                 -macTrace ON \
                 -movementTrace OFF

for {set i 0} {$i < $val(nn) } {incr i} {
     set node_($i) [$ns_ node]
     $node_($i) random-motion 0;
}

# ==============================================
# Set Node Positions
# ==============================================
$node_(0) set X_ 5.0
$node_(0) set Y_ 5.0
$node_(0) set Z_ 0.5

$node_(1) set X_ 6.0
$node_(1) set Y_ 5.0
$node_(1) set Z_ 0.5

$node_(2) set X_ 5.5
$node_(2) set Y_ 5.0
$node_(2) set Z_ 0.5
```

```
$ns_ at 0.10 "$node_(0) setdest 5.0 5.0 0.50"
$ns_ at 0.10 "$node_(1) setdest 6.0 5.0 0.50"
$ns_ at 0.10 "$node_(2) setdest 5.5 5.0 0.50"

#========================================
# Setup Traffic Flows
#========================================
set udp [new Agent/UDP]
$udp set fid_ 1
set sink [new Agent/LossMonitor]

$ns_ attach-agent $node_(0) $udp
$ns_ attach-agent $node_(1) $sink

$ns_ connect $udp $sink

# Creating CBR Traffic
set cbr [new Application/Traffic/CBR]
$cbr set packetSize_ 1
$cbr set interval_ 10.0
$cbr attach-agent $udp
$ns_ at 0.0 "$cbr start"

#========================================
# Simulation Startup and Shutdown
#========================================
for {set i 0} {$i < $val(nn) } {incr i} {
    $ns_ at 250.0 "$node_($i) reset";
}
$ns_ at 250.0 "stop"
$ns_ at 250.50 "puts \"NS EXITING...\" ; $ns_ halt"
proc stop {} {
    global ns_ tracefd nf
    $ns_ flush-trace
    close $tracefd
    close $nf
    exit 0
}
puts "Starting Simulation..."
$ns_ run
```

Mobility of underwater nodes creates problems which are unique to such a network. This makes it important to be able to simulate such scenarios as well.

APPENDIX C. NS-2 SAMPLE SCRIPTS

The script shown in this section extends the script from the previous section by making the nodes mobile. In order to achieve node mobility, within the "Set Node Positions" section instructions are added for NS-2 to move nodes to a certain position. While this mobility model is not complex, it is simple enough to demonstrate how mobility may also be built in to NS-2 scripts. If more complex mobility patterns are needed then the mobility patterns may be generated by using the NS-2 scenario tools. The files generated by these tools can be used within the AquaTools scripts just as they would be used in any other NS-2 simulation script.

C.3 Sample 3 - Energy Model

```
# ==========================================
# Define options
# ==========================================
set val(chan)           Channel/UnderwaterChannel
set val(prop)           Propagation/UnderwaterThorp
set val(netif)          Phy/UnderwaterPhy
set val(mac)            Mac/802_11
set val(ifq)            CMUPriQueue
set val(ll)             LL
set val(ant)            Antenna/OmniAntenna
set val(ifqlen)         50
set val(nn)             3
set val(rp)             DSR

set opt(energymodel)    EnergyModel
set opt(initialenergy)  97.0
set rx 0.75
set tx 2.0


Phy/UnderwaterPhy set CPThresh_ 10.0;
Phy/UnderwaterPhy set CSThresh_ 0.284;
Phy/UnderwaterPhy set RXThresh_ 4.0;
Phy/UnderwaterPhy set Pt_ 97;
Phy/UnderwaterPhy set freq_ 30;
Phy/UnderwaterPhy set L_ 1.0;



# ==========================================
# Setup Trace
# ==========================================
set ns_             [new Simulator]
```

APPENDIX C. NS-2 SAMPLE SCRIPTS

```
set tracefd       [open underwatertrace.tr w]
set nf [open nam-simple.nam w]
$ns_ use-newtrace ;
$ns_ trace-all $tracefd
$ns_ namtrace-all-wireless $nf 15 15


#============================================
# Setup Topography and Create God Process
#============================================
set topo          [new Topography]
$topo load_flatgrid 15 15
create-god $val(nn)


#============================================
# Create Nodes
#============================================
$ns_ node-config -adhocRouting $val(rp) \
                -llType $val(ll) \
                -macType $val(mac) \
                -ifqType $val(ifq) \
                -ifqLen $val(ifqlen) \
                -antType $val(ant) \
                -propType $val(prop) \
                -phyType $val(netif) \
                -channelType $val(chan) \
                -energyModel $opt(energymodel) \
                -rxPower $rx \
                -txPower $tx \
                -initialEnergy $opt(initialenergy) \
                -topoInstance $topo \
                -agentTrace ON \
                -routerTrace ON \
                -macTrace ON \
                -movementTrace OFF

for {set i 0} {$i < $val(nn) } {incr i} {
      set node_($i) [$ns_ node]
      $node_($i) random-motion 0;
}



#============================================
```

APPENDIX C. NS-2 SAMPLE SCRIPTS

```
# Set Node Positions
# ================================
$node_(0) set X_ 5.0
$node_(0) set Y_ 5.0
$node_(0) set Z_ 0.5

$node_(1) set X_ 6.0
$node_(1) set Y_ 5.0
$node_(1) set Z_ 0.5

$node_(2) set X_ 5.5
$node_(2) set Y_ 5.0
$node_(2) set Z_ 0.5

# ================================
# Setup Traffic Flows
# ================================
set udp [new Agent/UDP]
$udp set fid_ 1
set sink [new Agent/LossMonitor]

$ns_ attach-agent $node_(0) $udp
$ns_ attach-agent $node_(1) $sink

$ns_ connect $udp $sink

# Creating CBR Traffic
set cbr [new Application/Traffic/CBR]
$cbr set packetSize_ 1
$cbr set interval_ 10.0
$cbr attach-agent $udp
$ns_ at 0.0 "$cbr start"

# ================================
# Simulation Startup and Shutdown
# ================================
for {set i 0} {$i < $val(nn) } {incr i} {
    $ns_ at 250.0 "$node_($i) reset";
}
$ns_ at 250.0 "stop"
$ns_ at 250.50 "puts \"NS EXITING...\" ; $ns_ halt"
proc stop {} {
    global ns_ tracefd nf
```

```
        $ns_ flush-trace
        close $tracefd
        close $nf
        exit 0
}
puts "Starting Simulation..."
$ns_ run
```

Underwater acoustic networks have access to limited energy resources since the batteries installed on nodes are not very efficient and cannot be recharged while the network is deployed. As such, monitoring energy consumption of a network is an important aspect. NS-2 provides an energy model which can keep track of energy consumption in a network as well. Since AquaTools is able to utilize all features of NS-2, it can also track energy usage by using the same energy model. The script in this section shows how to utilize this energy model.

In the "Define Options" section the NS-2 energy model is chosen along with the initialization energy specified in Joules. The transmission and reception energy consumption values are also specified here in mW. Once these parameters are chosen, they need to be applied to the nodes as well. This can be achieved in the node configuration procedure. A sample of this can be seen in the "Configure Node" section.

These scripts show how any NS-2 feature can be used by the AquaTools toolkit. Any further complex scripts can be built easily in order to design more complex network designs and also test other routing protocols, MAC layers or energy saving schemes as well.

Acronyms

AN attenuation noise

AUV autonomous underwater vehicle

DSSS direct sequence spread spectrum

DTN disruption-tolerant network

FHSS frequency hopping spread spectrum

GPS global positioning system

ICoN integrated communication and navigation

IP internet protocol

MAC media access control

MSN mobile sensor node

NS-2 network simulator version 2

RF radio frequency

SNR signal-to-noise ratio

TCP transmission control protocol

USARSim urban search and resucue simulator

USN underwater sensor node

UUV unmanned underwater vehicle

UW-ASN underwater acoustic sensor network

UWSN underwater wireless sensor network

WSN wireless sensor network

WSS wireless simulation server

References

[1] I. F. Akyildiz, D. Pompili, and T. Melodia, "Underwater acoustic sensor networks: research challenges," *Ad Hoc Networks (Elsevier)*, vol. 3, pp. 257–279, 2005.

[2] I. F. Akyildiz, D. Pompili, and T. Melodia, "Challenges for efficient communication in underwater acoustic sensor networks," *SIGBED Rev.*, vol. 1, no. 2, pp. 3–8, 2004.

[3] Z. Peng, J. Cui, B. Wang, K. Ball, and L. Freitag, "An underwater network testbed: design, implementation and measurement," in *WuWNet '07: Proceedings of the second workshop on underwater networks*, (New York, NY, USA), pp. 65–72, ACM, 2007.

[4] J. Proakis, E. Sozer, J. Rice, and M. Stojanovic, "Shallow water acoustic networks," *IEEE Communications Magazine*, vol. 39, no. 11, pp. 114–119, 2001.

[5] E. Sozer, J. Proakis, R. Stojanovic, J. Rice, A. Benson, and M. Hatch, "Direct sequence spread spectrum based modem for under water acoustic communication and channel measurements," *OCEANS '99 MTS/IEEE. Riding the Crest into the 21st Century*, vol. 1, pp. 228–233, 1999.

[6] J. Partan, J. Kurose, and B. N. Levine, "A survey of practical issues in underwater networks," in *WUWNet '06: Proceedings of the 1st ACM international workshop on underwater networks*, (New York, NY, USA), pp. 17–24, ACM, 2006.

[7] A. F. Harris and M. Zorzi, "Modeling the underwater acoustic channel in ns2," in *ValueTools '07: Proceedings of the 2nd international conference on Performance evaluation methodologies and tools*, (ICST, Brussels, Belgium, Belgium), pp. 1–8, ICST (Institute for Computer Sciences, Social-Informatics and Telecommunications Engineering), 2007.

[8] M. Stojanovic, "On the relationship between capacity and distance in an underwater acoustic communication channel," in *WUWNet '06: Proceedings of the 1st ACM international workshop on underwater networks*, (New York, NY, USA), pp. 41–47, ACM, 2006.

REFERENCES

[9] K. V. MacKenzie, "Discussion of sea water sound-speed determinations," *Acoustical Society of America Journal*, vol. 70, pp. 801–806, Sept. 1981.

[10] C. C. Leroy, "Development of Simple Equations for Accurate and More Realistic Calculation of the Speed of Sound in Seawater," *Acoustical Society of America Journal*, vol. 46, pp. 216–+, 1969.

[11] K. V. MacKenzie, "Nine-term equation for sound speed in the oceans," *Acoustical Society of America Journal*, vol. 70, pp. 807–812, Sept. 1981.

[12] M. Herman, "Speed of sound in water for realistic parameters," *Acoustical Society of America Journal*, vol. 58, p. 1318, 1975.

[13] W. D. Wilson, "Equation for the Speed of Sound in Sea Water," *Acoustical Society of America Journal*, vol. 32, pp. 1357–+, 1960.

[14] H. G. Urban, *Handbook of Underwater Acoustic Engineering*. STN ATLAS Elektronik GmbH, November 2002.

[15] R. J. Urick, *Principles of Underwater Sound*. Los Altos, California: Peninsula Publishing, third ed., 1983.

[16] W. F. Baker, "New formula for calculating acoustic propagation loss in a surface duct in the sea," *Acoustical Society of America Journal*, vol. 57, pp. 1198–1200, May 1975.

[17] W. H. Thorp, "Analytic Description of the Low-Frequency Attenuation Coefficient," *Acoustical Society of America Journal*, vol. 42, pp. 270–+, 1967.

[18] W. H. Thorp, "Deep-Ocean Sound Attenuation in the Sub- and Low-Kilocycle-per-Second Region," *Acoustical Society of America Journal*, vol. 38, pp. 648–+, 1965.

[19] F. H. Fisher and V. P. Simmons, "Sound absorption in sea water," *The Journal of the Acoustical Society of America*, vol. 62, no. 3, pp. 558–564, 1977.

[20] M. A. Ainslie and J. G. McColm, "A simplified formula for viscous and chemical absorption in sea water," *Acoustical Society of America Journal*, vol. 103, pp. 1671–1672, Mar. 1998.

[21] M. Ali, U. Saif, A. Dunkels, T. Voigt, K. Römer, K. Langendoen, J. Polastre, and Z. A. Uzmi, "Medium access control issues in sensor networks," *SIGCOMM Comput. Commun. Rev.*, vol. 36, no. 2, pp. 33–36, 2006.

[22] J. hong Cui, J. Kong, M. Gerla, and S. Zhou, "Challenges: Building scalable mobile underwater wireless sensor networks for aquatic applications," in *IEEE Network, Special Issue on Wireless Sensor Networking*, pp. 12–18, 2006.

REFERENCES

[23] J. Heidemann, W. Ye, J. Wills, A. Syed, and Y. Li, "Research challenges and applications for underwater sensor networking," in *IEEE Wireless Communication and Networking Conference*, April 2006.

[24] E. Sozer, M. Stojanovic, and J. Proakis, "Underwater acoustic networks," *IEEE Journal of Oceanic Engineering*, vol. 25, pp. 72–83, January 2000.

[25] E. M. Sozer, M. Stojaovic, J. G. Proakis, J. A. Rice, M. Hatch, and A. Benson, "Direct sequence spread spectrum based modem for underwater acoustic communication and channel measurements," in *Proc. of the OCEANS Conference*, vol. 1, pp. 228–233, IEEE/MTS, 1999.

[26] L. Freitag, M. Stojanovic, S. Singh, and M. Johnson, "Analysis of channel effects on direct-sequence and frequency-hopped spread-spectrum acoustic communications," *IEEE Journal of Oceanic Engineering*, vol. 26, no. 4, pp. 586–593, 2001.

[27] D. Kalofonos, M. Stojanovic, and J. Proakis, "Performance of adaptive mc-cdma detectors in rapidly fading rayleigh channels," *IEEE Transactions on Wireless Communications*, vol. 2, no. 2, pp. 229–239, 2003.

[28] V. Bharghavan, A. Demers, S. Shenker, and L. Zhang, "Macaw: a media access protocol for wireless lan's," *SIGCOMM Comput. Commun. Rev.*, vol. 24, no. 4, pp. 212–225, 1994.

[29] J. Rice, B. Creber, C. Fletcher, P. Baxley, K. Rogers, K. McDonald, D. Rees, M. Wolf, S. Merriam, R. Mehio, J. Proakis, K. Scussel, D. Porta, J. Baker, J. Hardiman, and D. Green, "Evolution of seaweb underwater acoustic networking," in *Proc. OCEANS 2000 MTS/IEEE Conference and Exhibition*, vol. 3, pp. 2007–2017, 11–14 Sept. 2000.

[30] J. Rice, "Seaweb acoustic communication and navigation networks," July 2005.

[31] L. Freitag, M. Grund, C. von Alt, R. Stokey, and T. Austin, "A shallow water acoustic network for mine countermeasures operations with autonomous underwater vehicles," *Underwater Defense Technology (UDT)*, 2005.

[32] G. Acar and A. E. Adams, "Acmenet: an underwater acoustic sensor network protocol for real-time environmental monitoring in coastal areas," *IEE Proceedings -Radar, Sonar and Navigation*, vol. 153, pp. 365–380, August 2006.

[33] M. Molins and M. Stojanovic, "Slotted fama: a mac protocol for underwater acoustic networks," in *Proc. OCEANS 2006 - Asia Pacific*, pp. 1–7, 16–19 May 2007.

REFERENCES

[34] B. Peleato and M. Stojanovic, "A mac protocol for ad-hoc underwater acoustic sensor networks," in *WUWNet '06: Proceedings of the 1st ACM international workshop on Underwater networks*, (New York, NY, USA), pp. 113–115, ACM, 2006.

[35] F. Schill, U. R. Zimmer, and J. Trumpf, "Towards optimal tdma scheduling for robotic swarm communication," in *Proceedings of the TAROS (Towards Autonomous Robotic Systems) intl. conference*, September 2005.

[36] M. Dunbabin, P. Corke, I. Vasilescu, and D. Rus, "Data muling over underwater wireless sensor networks using an autonomous underwater vehicle," in *Proc. IEEE International Conference on Robotics and Automation ICRA 2006*, pp. 2091–2098, 15–19 May 2006.

[37] W. Zhao and M. H. Ammar, "Message ferrying: proactive routing in highly-partitioned wireless ad hoc networks," in *Proc. Ninth IEEE Workshop on Future Trends of Distributed Computing Systems FTDCS 2003*, pp. 308–314, 28–30 May 2003.

[38] W. Zhao, M. Ammar, and E. Zegura, "A message ferrying approach for data delivery in sparse mobile ad hoc networks," in *MobiHoc '04: Proceedings of the 5th ACM international symposium on Mobile ad hoc networking and computing*, (New York, NY, USA), pp. 187–198, ACM, 2004.

[39] B. Burns, O. Brock, and B. N. Levine, "Autonomous enhancement of disruption tolerant networks," in *Proc. IEEE International Conference on Robotics and Automation ICRA 2006*, pp. 2105–2110, 15–19 May 2006.

[40] B. Burns, O. Brock, and B. N. Levine, "Mora routing and capacity building in disruption-tolerant networks," *Ad Hoc Netw.*, vol. 6, no. 4, pp. 600–620, 2008.

[41] L. Freitag, M. Johnson, M. Grund, S. Singh, and J. Preisig, "Integrated acoustic communication and navigation for multiple uuvs," in *Proc. MTS/IEEE Conference and Exhibition OCEANS*, vol. 4, pp. 2065–2070, 5–8 Nov. 2001.

[42] L. E. Freitag, M. Grund, J. Partan, K. Ball, S. Singh, and P. Koski, "Multi-band acoustic modem for the communications and navigation aid auv," in *Proc. MTS/IEEE OCEANS*, pp. 1080–1085, 17–23 Sept. 2005.

[43] M. Stojanovic, L. Freitag, J. Leonard, and P. Newman, "A network protocol for multiple auv localization," in *Proc. Oceans '02 MTS/IEEE*, vol. 1, pp. 604–611, 29–31 Oct. 2002.

[44] R. R. Kanthan, "The icon integrated communication and navigation protocol for underwater acoustic networks," Master's thesis, MIT, September 2005.

REFERENCES

[45] E. M. Sozer, M. Stojanovic, and J. G. Proakis, "Design and simulation of an underwater acoustic local area network," in *Proc. Opnetwork 99*, 1999.

[46] K. P. Prasanth, "Modelling and simulation of an underwater acoustic communication channel," Master's thesis, Hochschule Bremen, University of Applied Sciences, August 2004.

[47] F. Vanni, A. Aguiar, and A. Pascoal, "Networked marine systems simulator." May 2008.

[48] Z. Peng, J.-H. Cui, B. Wang, K. Ball, and L. Freitag, "An underwater network testbed: design, implementation and measurement," in *WuWNet '07: Proceedings of the second workshop on underwater networks*, (New York, NY, USA), pp. 65–72, ACM, 2007.

[49] R. Coates, *Underwater Acoustic Systems*. Wiley, 1989.

[50] S. Shell, P. Debenedetti, and A. Panagiotopoulos, "Molecular structural order and anomalies in liquid silica," *Phys. Rev. E Stat. Nonlin. Soft. Matter. Phys*, vol. 66, 2002.

[51] NS-2, "The network simulator manual," *http://www.isi.edu/nsnam/ns/*, 28 July 2009.

[52] L. Freitag, M. Grund, S. Singh, J. Partan, P. Koski, and K. Ball, "The whoi micro-modem: An acoustic communications and navigation system for multiple platforms," *http://www.whoi.edu*, 2005.

[53] A. Sehgal, I. Tumar, and J. Schoenwaelder, "Variability of available capacity due to the effects of depth and temperature in the underwater acoustic communication channel," in *Proc. of IEEE OCEANS '09*, (Bremen), May 2009.

[54] UCAR, "Temperature of ocean water," *http://www.windows.ucar.edu/tour/link=/earth/Water/temp.html&edu=high*, 31 August 2001.

[55] A. Caiti, E. Crisostomi, and A. Munafo, "Physical characterization of acoustic communication channel properties in underwater mobile sensor networks," in *Proceedings of International Conference on Sensor Systems and Software*, (Pisa, Italy), September 2009.

[56] UCAR, "Standard salinity profile," *http://www.windows.ucar.edu/tour/link=/earth/Water/salinity_depth.html&edu=high*, 31 August 2001.

[57] H. Sverdrup, M. W. Johnson, and R. H. Fleming, *The World's Oceans: Their Physics, Chemistry, and General Biology*. Englewood Cliffs, New Jersey: Prentice Hall Inc., 1970.

www.ingramcontent.com/pod-product-compliance
Lightning Source LLC
Chambersburg PA
CBHW080932170526
45158CB00008B/2262